BIOTECHNOLOGY
AND
GENETIC ENGINEERING

LIBRARY IN A BOOK

BIOTECHNOLOGY
AND
GENETIC ENGINEERING

Lisa Yount

☑®
Facts On File, Inc.

BIOTECHNOLOGY AND GENETIC ENGINEERING

Facts On File, Inc.
11 Penn Plaza
New York NY 10001

Library of Congress Cataloging-in-Publication Data

Yount, Lisa. Biotechnology and genetic engineering / Lisa Yount.
    p. cm.—(Library in a book)
Includes bibliographical references and index.
ISBN 0-8160-4000-1 (hc : acid-free paper)
    1. Biotechnology—Social aspects. 2. Genetic engineering—Social aspects.
I. Title. II. Series.
TP248.23 .Y684 2000
303. 48'3—dc21   99–049532

Facts On File books are available at special discounts when purchased in bulk quantities for businesses, associations, institutions, or sales promotions. Please call our Special Sales Department in New York at 212/967-8800 or 800/322-8755.

You can find Facts On File on the World Wide Web at
http://www.factsonfile.com

Text design by Ron Monteleone

Printed in the United States of America

MP FOF 10 9 8 7 6 5 4 3 2

This book is printed on acid-free paper.

To the scientists

who will create the "Biotech Century"

and the children

who will have to deal with it

CONTENTS

PART III
APPENDICES

PART I

OVERVIEW OF THE TOPIC

CHAPTER 1

ISSUES IN BIOTECHNOLOGY AND GENETIC ENGINEERING

In some senses, biotechnology—the use or alteration of other living things in processes that benefit humankind—is almost as old as humanity itself. Biotechnology certainly has existed as long as people have raised crops and domesticated animals, practices they began some 10,000 years ago. Although they were unaware that they were doing so, people used microorganisms in the processes of making bread, cheese, alcoholic drinks, and tanned leather. When people began sowing only the seeds from the best (most productive, most disease-resistant, and so on) of the previous year's crop plants, and breeding only the strongest or most productive cattle and other domestic animals, they also, again unknowingly, altered the genes of living things to better meet their needs. When they learned that they could breed certain plants from cuttings, they discovered cloning (the word *clone* means "twig" in Greek).

CRACKING THE CODE OF LIFE

Only in the 19th century did people begin to learn what was really happening during these technological efforts. Starting in the late 1850s, French chemist Louis Pasteur showed that living yeast or other microorganisms were necessary for the fermentation that produced wine, beer, and other products. More important to the genetic side of modern biotechnology, in 1866 an obscure Austrian monk named Gregor Mendel published a paper describing the mathematical rules of inheritance of physical traits. He had worked out these rules by breeding and observing pea plants in his monastery garden. Mendel's rules showed the statistical pattern in which

3

such characteristics as height and seed color were inherited by the hybrid off-spring of two plants that differed from each other in these characteristics.

The form of a characteristic that an offspring would receive was determined by what Mendel called factors, one of which (for each characteristic) the offspring inherited from its male parent and one from its female parent. Some factors were more powerful, or dominant, than others. If a plant inherited one dominant factor and one weaker (recessive) factor for a trait, the plant would exhibit the dominant form of the trait. Few knew of Mendel's work at the time, let alone realized that it went far toward explaining the theory of Charles Darwin, described in his *On the Origin of Species* (first published in 1859), which stated that nature acted like traditional plant and animal breeders to select and preserve the inherited characteristics that helped species survive in their environment.

Earlier in the 19th century, Matthias Schleiden and Theodore Schwann had proposed that the microscopic bodies called cells were the basic units from which all living things were formed. Improvements in microscopes allowed scientists to begin exploring the cell's inner structures toward the end of the century. Mendel's rules of heredity and the study of cells converged in the early 1900s, when Mendel's forgotten work was rediscovered and scientists began speculating that his "factors" were somehow contained in threadlike chromosomes, or "colored bodies," which German biologist Walther Flemming had discovered in the nuclei (central bodies) of cells in 1875. (All cells except those of bacteria and other so-called prokaryotes contain nuclei and chromosomes.)

Drawing on experiments with fruit flies, Thomas Hunt Morgan of Columbia University and his coworkers proved the link between chromosomes and heredity in 1910. By then Mendel's factors had taken on the new name of *genes*, and the science of genetics had been born. In the decades that followed, Morgan's group and others began to make maps of chromosomes showing approximately where the genes that determined certain characteristics were located.

Geneticists assumed that inherited information must be coded somehow in the structure of either proteins or nucleic acids, the only types of chemical in the chromosomes. At first they thought proteins the more likely information carriers because these chemicals were known to have a complex structure. In 1944, however, Oswald Avery and his coworkers showed that a harmless strain of certain bacteria became capable of causing disease when they took up deoxyribonucleic acid (DNA) from a disease-causing strain of the same bacteria. The bacteria's descendants retained the new disease-causing ability, indicating that information for producing this trait must have been contained in the DNA. Avery's work turned the genetic spotlight from proteins to nucleic acids, especially DNA.

Issues in Biotechnology and Genetic Engineering

At that time DNA's chemical composition was known, but key aspects of its structure were not. Biologists knew that the DNA molecule had a long "backbone" composed of smaller molecules of phosphate and a sugar called deoxyribose. Attached to this backbone were many units of four kinds of small molecules called bases: adenine (A), thymine (T), guanine (G), and cytosine (C). X-ray crystallography photographs suggested that the DNA molecule had the general shape of a helix, or coil. However, no one knew exactly how many backbone chains the molecule had, how the bases were attached to them, or how all these units were packed within the molecule.

Geneticists and biochemists in the late 1940s and early 1950s increasingly realized that the structure of DNA must hold the secret of its ability to transmit biological instructions from one generation to the next, and several groups of scientists began trying to figure out that structure. The ones who first succeeded were a young American, James Watson, and a somewhat older Englishman, Francis Crick, who worked together at England's prestigious Cambridge University. Drawing on molecular models and X-ray photographs, they concluded that each DNA molecule contained two sugar-phosphate backbones that entwined one another. The shape this formed came to be known as the "double helix." Pairs of bases stretched between the two backbones like rungs on a twisted ladder. A molecule of adenine always paired with one of thymine and, similarly, cytosine paired with guanine.

Watson and Crick published their groundbreaking account of DNA's structure in the British science journal *Nature* on April 25, 1953. Five weeks later, they published a second paper describing how this structure could allow DNA to reproduce itself so that a complete copy of hereditary instructions was given to both of the "daughter" cells when a cell divided. The hydrogen bonds that joined the bases in each pair were weak, Watson and Crick explained, so the bonds might dissolve as the cell prepared to divide. Each DNA molecule would thus split down its length, similar to a zipper unzipping. Each base could then attract a molecule of the complementary base, bearing with it a sugar-phosphate "backbone" unit, from free-floating materials in the cell, and the bonds joining the bases would re-form. The result would be two separate DNA molecules identical to the original one. This theory was proved correct in 1958.

DNA's structure also revealed the "code" in which inherited information is stored. Molecular biologists had known since 1941 that genes contain instructions for making proteins, the complex chemicals that do most of the work of the cell. Indeed, scientists had come to define a gene as the portion of a DNA molecule that carries the instructions for making one protein. (This proved to be an oversimplification. Some genes make molecules of ribonucleic acid [RNA] or modify the activity of other genes.) The most likely candidate for the code in which these instructions were carried was the order of the bases in the DNA molecule. DNA has only four kinds of bases, however, whereas proteins consist of 20 types of simpler substances called amino acids. In 1961, Crick suggested that the "code letter" specifying a particular type of amino acid could be a sequence of three bases. The 64 (4 x 4 x 4) possible triads were more than enough to specify all of the amino acids.

During the next five years, several groups of scientists painstakingly deciphered this genetic code, determining which amino acid each of the 64 triads stood for. (Some amino acids could be represented by any of several different triads.) Meanwhile, others worked out the procedure by which information in the DNA code was translated into protein. They discovered that the DNA was copied into a form of RNA, a related chemical that differed from DNA only in having uracil rather than of thymine as one of its four bases. Unlike DNA, RNA could leave the cell's nucleus and move out into the cytoplasm, the main body substance of the cell. There the RNA guided the assembly of amino acids, in the order specified by the original DNA, into a protein molecule.

THE BIRTH OF GENETIC ENGINEERING

Building on these basic discoveries, scientists in the early 1970s developed the ability to change genes instead of merely deciphering them. Genetic engineering—the direct alteration of genes or transfer of genes from one type of organism to another—works because (except in the case of a few viruses with genes made of RNA rather than DNA) all genes have the same basic structure and work in the same way. Thus, they are in a sense interchangeable. Once a gene is placed in a cell's genome, or collection of genes, that cell becomes able to make the protein for which the gene codes, regardless of whether it ever made that substance in nature. As an example, bacteria or cattle can be made to produce human hormones by inserting the appropriate human genes into them.

Genetic engineering began in the laboratory of Paul Berg, a Stanford University biochemist, in the winter of 1972–73. Berg removed a gene from SV40 (Simian Virus 40), a monkey virus that could cause cancer in mice. Through laborious chemical means, he attached to it a short piece of single-stranded DNA. He then opened up the small, circular genome of another virus, lambda, and attached a chain of single-stranded DNA with a base sequence opposite to that on the added SV40 piece to one of the lambda genome's open ends. He termed these single-stranded chains "sticky ends" because any such chain will attach itself to another strand with a complementary (opposite) base sequence. (For example, a chain with the sequence A-A-T-T will attach itself to another with the sequence T-T-A-A.) Taking advantage of this "stickiness," Berg spliced the gene from SV40 into the lambda genome. This was the first production of recombinant DNA, or DNA into which other DNA from a different species of organism has been inserted.

Berg's technique for gene splicing would have been hard to apply on a mass scale, but the same was not true of another technique developed at roughly the same time by another Stanford scientist, Stanley Cohen, and Herbert Boyer, from the University of California at San Francisco. Appropriately, the idea for this technique—which would allow researchers to in effect slice and dice genes, sandwich them in any order, and pack them "to go"—was born in a delicatessen. Cohen and Boyer met at this delicatessen one evening in November 1972 after attending a scientific meeting in Honolulu, Hawaii. As they discussed their work, each discovered that the other's research held a missing piece of his own puzzle.

Unlike Berg, Cohen and Boyer were working with bacteria. A bacterium's genome is large compared to that of a virus, containing an average of 2,500 genes rather than a few hundred, although it is far smaller than those of cells with nuclei (the human genome, for instance, may contain up to about 150,000 genes). Most of a bacterium's genes are carried in a single large ring of DNA, but some are contained in smaller rings called plasmids, which the bacteria can exchange in several ways. Cohen had invented a technique for removing plasmids from one bacterial cell and inserting them into another.

Boyer, for his part, was working with bacteria called *Escherichia coli* (*E. coli* for short), which commonly and usually harmlessly live in the human intestine. He was studying enzymes (proteins that catalyze chemical reactions in living cells) called restriction endonucleases, which *E. coli* and some other bacteria produce as a defense against viruses. These enzymes snip DNA into pieces. Each restriction enzyme breaks a DNA molecule apart at every spot where it encounters a particular sequence of bases, and the process conveniently produces the same kind of single-stranded "sticky ends" that Berg had so painstakingly created with his chemicals. Any two pieces of DNA cut by the same restriction enzyme, therefore, can be joined together, even if they come from very different species. Different restriction enzymes act on different base sequences. More than 800 restriction enzymes are known today, and have become basic tools of gene splicing. So have the ligases, a type of enzymes that both Boyer and Berg had used to weld together their spliced DNA fragments.

Chatting over their corned beef sandwiches, Cohen and Boyer realized that if they applied Boyer's *E. coli* restriction enzyme, EcoR1, to Cohen's plasmids, they would have a way to cut open a plasmid from one species of bacteria and, with the help of a ligase, splice it onto a plasmid from another species. They proceeded to try this in spring 1973. One of their two plasmids contained a gene that conferred resistance to a certain antibiotic. Boyer and Cohen allowed a type of bacterium that was not naturally resistant to that antibiotic to take up the combined plasmids, and then placed the bacteria in a culture medium containing the antibiotic. Some of the bacteria survived, which showed that the resistance gene in the engineered plasmids was still making its signature protein. Later that year, the two scientists spliced some DNA from a frog into an *E. coli* plasmid and proved that it functioned there as well.

In addition to plasmids, Boyer, Cohen, and other scientists were soon using viruses (such as lambda, which infects bacteria) as vectors, or transmission agents, for inserting foreign genes into bacteria and, later, plant and animal cells. Viruses reproduce by inserting their genes into the genomes of cells, which then reproduce the viral genomes along with their own; when scientists add genes to the viruses, these are inserted and reproduced as well. As one scientist commented when he heard about the new techniques, "Now we can put together any DNA we want to."[1] The Gene Age had begun.

Although traditional biotechnology continues today, the term has now become almost synonymous with genetic engineering. Together, biotechnology and genetic engineering represent not only a scientific field but a burgeoning industry. In the two and a half decades since the "gene deli" opened, companies based on the technology of decoding and altering genes have multiplied seemingly as fast as the bacteria in their founders' test tubes. Although enthusiasm about the profitability of such companies has repeatedly waxed and waned, few people doubt that sooner or later genetic engineering and biotechnology will effect massive changes in the way people live, their health, their environment, and, perhaps, in the very nature of humanity itself. Supporters and critics alike expect the 21st century to be what one critic, Jeremy Rifkin, has dubbed "the biotech century."

When scientists and businesspeople began to tinker with the very stuff of life, their work raised serious ethical, legal, and social issues. This book focuses on five areas in which such issues are especially sensitive: health and environmental safety, particularly in regard to agricultural biotechnology; the patenting of living things, body parts or cells, and genes; DNA testing for identification in criminal cases; genetic testing for susceptibility to disease; and alteration of human genes. The remainder of this introduction will describe the issues in these areas, the technologies that spawned them, and the way governments, courts, and the public have perceived and reacted to them.

AGRICULTURAL BIOTECHNOLOGY: SAFETY ISSUES

Concerns about the safety of genetic engineering experiments and genetically altered organisms began even before genetic engineering itself. In 1971, when Robert Pollack, a geneticist at Cold Spring Harbor Laboratory on Long Island, New York, learned that Paul Berg planned to allow lambda containing SV40 genes to infect *E. coli*, he telephoned Berg to express his concern over the possible consequences of inserting genes from a cancer-causing virus into one that could infect bacteria that live in the human intestine. It might be possible, Pollack said, for bacteria containing cancer genes to escape from the laboratory and infect people. Berg decided that Pollack was right. He transferred genes from SV40 to lambda, but he did not allow the altered lambda to infect bacteria.

Other scientists, including Boyer and Cohen, came to share Berg's and Pollack's uneasiness. As some of them pointed out in a discussion at the Gordon Conference on Nucleic Acids in mid-1973, *E. coli*, the "workhorse" bacterium that many experimenters in the new field used, not only could infect humans itself but could exchange plasmids with other bacteria, including ones that cause human disease. The result of this and similar discussions was two letters to *Science*, the prestigious journal of the American Association for the Advancement of Science, warning of the possible dangers of recombinant DNA research. The second letter, signed by top scientists in the field including Berg, Boyer, and Cohen, and published in the July 26, 1974 issue of *Science*, was the more detailed of the two. It recommended halting several types of recombinant experiments, including those involving genes for antibiotic resistance or tumor formation, until possible hazards could be more thoroughly evaluated and guidelines for safe procedure could be established. This was the first time a group of scientists had voluntarily proposed halting a certain type of experiment because of possible dangers.

Fear of possible harm from accidental release of genetically engineered microorganisms brought 140 geneticists and molecular biologists together for a historic conference at Asilomar, a seaside retreat center in central California, in February 1975. Michael Rogers, a journalist writing in *Rolling Stone*, called it "the Pandora's Box Congress."[2] The group drew up guidelines that divided gene-splicing experiments into four categories. Category P1 experiments presented minimal risk to humans and required no precautions beyond those in normal laboratory practice, whereas those in category P4 potentially were highly dangerous. The Asilomar scientists concluded that "most of the work on construction of recombinant DNA molecules should proceed, provided that appropriate safeguards . . . are employed," but they agreed not to perform any experiments in the P3 or P4 categories until new

9

laboratories with special safeguards for containing potentially dangerous microbes could be built.[3] The scientists also promised to modify the strain of *E. coli* that they used so that the germs would be unlikely to survive outside a laboratory.

The Asilomar guidelines became the blueprint for the first federal government regulations on genetic engineering. These regulations, issued in June 1976, were drafted by the Recombinant DNA Advisory Committee (RAC), a panel of molecular biologists and other scientists that the National Institutes of Health (NIH) had established in October 1974. The RAC had to approve all recombinant DNA experiments funded by the NIH, which at the time meant most of such experiments in the United States. The NIH regulations were binding on all federally funded research, and most private (usually university-based) researchers agreed to abide by them as well. About this same time, Britain published similar guidelines and established a government oversight group much like the RAC. The European Molecular Biology Organization (EMBO) recommended that researchers in other European countries follow either the US/NIH or the British guidelines.

Worry about the possible escape of genetically engineered "superbugs" was not limited to scientists. Congress held its first hearings on the issue in April 1975, only a few months after the Asilomar meeting took place. Sensationalist media articles, often blending fact with speculation, stirred up the public's fears. Critics, including several environmental groups, complained that the NIH guidelines were not strict enough, accused RAC scientists of conflict of interest (many of them carried out the kind of research they were supposed to regulate), and demanded that ethicists and environmentalists have a hand in drawing up future regulations. Debate about the new science was so intense in the mid-1970s that Rockefeller University professor Norton Zinder called it the "Recombinant DNA War."[4]

Friends and foes alike were concerned that the NIH had little, if any, legal authority to regulate research other than that which it funded. Indeed, a federal interagency committee reported to U.S. Health, Education, and Welfare Secretary Joseph Califano on March 15, 1977, that, as far as it could tell, no existing federal agency had the legal authority to handle all problems raised by recombinant DNA research. Some state and local groups tried to fill this void by passing their own regulations. In July 1976, for instance, the city council of Cambridge, Massachusetts—home to university giants Harvard and MIT—announced at the end of a highly publicized and disputatious meeting that they were imposing a three-month moratorium on recombinant DNA research in the city. New York and California both considered regulatory legislation.

In 1977, partly to prevent what they saw as usurpation of federal authority and imposition of a "patchwork" of state and local rules, members of

Issues in Biotechnology and Genetic Engineering

Congress introduced 16 different bills to regulate recombinant DNA research. None of the bills became law, however. By this time public attention had begun to wander elsewhere, and most scientists in the field had come to fear restrictive legislation even more than runaway bacteria and united to lobby against it. They convinced the legislators that the strain of *E. coli* used in most recombinant DNA experiments presented no threat to public health.

The scientists were so successful in easing governmental fears that the NIH relaxed its guidelines in 1978 and again in 1980, downgrading many experiments to lower containment levels and exempting others—eventually most—from regulation entirely. As the introduction to the 1978 revision noted,

> *No new facts or unconsidered older ones have emerged to support the fears of harmful effects [of recombinant DNA research]. . . . [T]here is growing sentiment that the burden of proof is shifting toward those who would restrict . . . [such] research.*[5]

Indeed, by the early 1980s most scientific and regulatory bodies had ceased to regard the process of genetic engineering as inherently risky. Regulatory attention shifted from gene-altering experiments to the organisms they produced.

Fears about the safety of genetically engineered organisms flared up again in the mid 1980s, when experimenters and fledgling biotechnology companies began planning to release such organisms into the environment. One of the first researchers to feel the heat was Steven Lindow, a plant pathologist at the University of California at Berkeley.

Lindow had discovered in 1975 that bacteria called *Pseudomonas syringae* contain a surface protein that encourages ice crystals to form on the bacteria at low temperatures. These ubiquitous bacteria live on plants, and their presence made crops such as citrus, strawberries, and potatoes susceptible to frost damage. Such damage cost United States farmers up to $1.5 billion a year. Around 1978, Lindow stumbled across a natural mutant of these bacteria that had lost the gene coding for the ice-forming protein. He and his coworkers used the mutants to identify the gene and then developed a process for removing it from normal *Pseudomonas* bacteria. The scientists demonstrated in greenhouse experiments that when the resulting "ice-minus" bacteria were sprayed on susceptible plants at the beginning of the cold season, they outcompeted the normal bacteria and kept plants free of frost at temperatures where heavy damage would ordinarily occur.

In 1982, Lindow and Advanced Genetic Sciences, a company formed to develop the ice-minus bacteria commercially, applied to the RAC for permission to test the altered bacteria in small open fields owned by the university. They received preliminary approval in June 1983. However, Jeremy Rifkin, director of the Foundation on Economic Trends and a foe of genetic engineering since the "recombinant DNA war" days, found out about the proposed test and began to drum up opposition to it, calling it "ecological roulette."[6] Environmental groups who joined Rifkin's protests feared that the altered bacteria might spread out of their test fields and change local ecologies or even alter rainfall patterns. Questions from the RAC and lawsuits from Rifkin and his followers delayed Lindow's tests for four years. When the tests were finally conducted—in spite of last-minute sabotage attempts— on April 24, 1987, this first release of genetically altered organisms into the environment produced no significant changes in either ecology or climate.

By this time, genetic engineering was being applied to plants and animals as well as bacteria, and the RAC was no longer the only federal body regulating it. In June 1986, the U.S. Office of Science and Technology Policy had published the Coordinated Framework for Regulation of Biotechnology, which divided regulation of genetic engineering research and products among five agencies: the NIH, the National Science Foundation (NSF), the U.S. Department of Agriculture (USDA), the Environmental Protection Agency (EPA), and the Food and Drug Administration (FDA). According to the framework, the first two groups would evaluate research supported by grants; the USDA specifically, its Animal and Plant Health Inspection Service (APHIS) would regulate genetically altered agricultural plants and animals; the FDA would cover engineered drugs, medical treatments, and foods; and the EPA would handle anything related to pesticides and other chemicals. In that same year, the Toxic Substances Control Act (TSCA) was amended to require a permit from the EPA to release most genetically engineered organisms into the environment, use them in manufacturing, or distribute them commercially for intended release. Lindow thus had to obtain a permit from the EPA as well as the RAC before carrying out his tests.

Most of the products that followed Lindow's lead into open-field testing were plants rather than bacteria. By 1994, over 2,000 field tests of transgenic plants (those containing genes from another species) had taken place worldwide. No obvious environmental disasters resulted from the tests, and regulation of field testing of genetically altered plants, like regulation of experiments on altered microorganisms before it, as a result slowly relaxed. In 1993, for instance, the USDA agreed to require only notification of the agency, not a permit, for testing of genetically altered strains of corn, cotton, potatoes, soybeans, tomatoes, and tobacco, which together represented 80 to 90 percent of applications. On the whole, regulatory agencies in the United States seem to agree with the claims of agricultural biotechnology companies that genetically engineered plants are not substantially different from those changed by traditional breeding techniques and therefore require no additional regulation.

Today, transgenic crops are a multibillion-dollar business. On either a commercial or an experimental basis, plants have been engineered to resist insects, disease, drought, salt, and herbicides and to produce everything from plastics to vaccines. By mid-1999, more than 50 genetically altered crops, including at least 24 food crops, had been approved for sale in the United States. In 1996, when transgenic crops were first grown on a commercial basis, only 4 million acres in the United States were planted with genetically altered seeds, but in 1998 the figure had risen to at least ten times that. Nearly half of the country's cotton, one-third of its soybeans, and 15 percent of its corn were genetically modified. Although Europe and some other parts of the world have resisted planting transgenic crops, a total of about 100 million acres worldwide were planted with such crops in 1999, including over half of the world's supply of soybeans and about one-third of its corn. That number was expected to triple in five years. In addition to the United States, countries with large transgenic crop acreage include Argentina, Canada, and China.

This is not to say that all safety concerns about genetically engineered plants have vanished. Critics say alterations that seem harmless in tests may damage the environment or affect human health when the crops carrying them become widespread. They claim that dependence on genetically engineered crops also reduces biological diversity, which is already seriously threatened by the worldwide tendency to plant large areas in genetically identical or near-identical natural crops. This narrow genetic base can turn a plant disease epidemic into a major disaster.

An example of plant genetic engineering that may have unexpected environmental effects is resistance to herbicides, which has been engineered into crop plants such as Monsanto Corporation's "Roundup Ready" soybeans, canola (an oil-producing crop), and corn. Agribusiness companies sell seeds for these crops and the herbicide to which they have been made resistant as a package. Indeed, they often require farmers to sign contracts promising to use only that company's brand of the chemical (Roundup, for example, is Monsanto's brand of the herbicide glyphosate).

Herbicide-resistant crops have become very popular in the United States. In 1999, for instance, about half of the country's soybean crop was expected to be Roundup Ready. These crops have somewhat higher yields than natural crops of the same kind, and they let farmers use herbicides more efficiently to control weeds. Farmers therefore need to use smaller amounts of the chemicals, supporters say. The crops can also free farmers from the need to plow, thus reducing loss of topsoil to erosion. In spite of the crops' high initial cost, their increased yield and decreased chemical requirements also save farmers money—$419 million in the United States and Canada in 1997, according to one estimate.

Environmentalists distrust the resistant crops. They fear that such crops will make farmers more likely to depend on and overuse herbicides, which could increase threats to farm workers' health and the likelihood that wild plants on which animals and birds depend for food will be destroyed. This latter concern is especially great in Europe, where farms and wild areas are often close together. Critics also cite evidence that herbicide-resistance genes can spread from crop plants to nearby weeds, which are often wild relatives of the crops. The weeds could then become resistant as well, leading to a new cycle of chemical dependence. The crop plants themselves may even escape from farmers' fields and become "superweeds."

Another popular type of genetically altered crop may also cause unexpected environmental damage. These crops have been given a gene from the bacterium *Bacillus thuringensis* (Bt for short) that codes for a natural insecticide. They therefore can produce their own pesticides. (Indeed, the EPA classifies the altered plants as pesticides.) Like herbicide-resistant crops, crops containing Bt have become commonplace on American farms. In 1999, for example, about 20 percent of the U.S. corn crop—some 8 million acres' worth—had added Bt genes. Farmers who plant these crops can greatly reduce their use of expensive, environment-damaging chemical pesticides.

Organic farmers have used sprays of Bt bacteria as "natural" pesticides for decades, but sunlight destroys the bacteria a few days after spraying. Crop plants containing Bt genes, by contrast, produce the bacterial toxin all the time. This means that pest insects are exposed to it for longer periods and thus have more chance to develop a resistance. Some evidence of increased Bt resistance in pest insects has already appeared.

Furthermore, although Bt spray normally does not harm nonpest insects, the same may not be true of the toxin produced by the plants. In May 1999, entomologists at Cornell University published the results of a laboratory study suggesting that pollen from Bt-engineered corn could be carried onto milkweed plants by wind and could then poison Monarch butterfly caterpillars feeding on those plants. Some environmentalists have called for a ban on Bt-engineered corn because of the study, but biotechnology industry representatives and others have questioned whether milkweed normally grows close enough to corn plants to receive much pollen. In addition, they say, possible harm caused by the pollen must be weighed against the plants' ability to reduce the use of chemical pesticides, which also kill butterflies and other nonpest insects.

Some critics say that genetically modified food crops pose threats to human health as well as to the environment. The first genetically engineered food, a type of tomato called Flavr Savr that had been altered to rot more slowly after ripening and thus be less liable to spoil during shipping and storage, was FDA-approved for sale in 1994. Ever since, groups have demanded the banning, or at least the labeling, of such "unnatural" foodstuffs. Defenders of the altered foods insist that most are nutritionally no different from their natural cousins and therefore need no special labeling or regulation. The FDA has agreed with this point of view, ruling in May 1992 that genetically altered foods do not need to be reviewed by the agency or labeled as long as they contained no new, toxic, or foreign substances that might cause health problems. A prestigious 1998 British government study also found no significant differences between altered and natural foods, but it recommended labeling genetically modified foods so that consumers could choose whether to eat them.

So far, the only proven health threat from genetically engineered foods relates to allergies. People allergic to Brazil nuts, for instance, also have allergic reactions to soybeans containing a gene from the nuts. Recognizing this potential problem, the FDA requires labels on products to which genes from known food allergens, such as peanuts, have been added. Other health problems may be starting to surface as well, however. Although Bt spray is considered harmless to human health, a few scientists have reported brain, liver, and immune system damage and possible cancer rate increases in rats that are fed Bt-engineered potato plants. The genetically engineered Bt toxin may have effects not evident in the sprayed version because the spray deteriorates quickly, whereas the engineered toxin is still in the plant when the plant is eaten. A 1999 study suggests that crop pickers and handlers exposed to Bt spray can develop allergies to it, so people eating Bt-containing plants might do so as well.

Protests against genetically altered foods in the late 1990s have been much stronger in Europe, where they are often called "Frankenfoods," than in the United States. A 1999 British poll reported that 77 percent of those surveyed wanted genetically modified crops banned. In 1997 the European Union began to require labeling of all foods that might contain genetically modified organisms and in 1999 it put at least a temporary ban on importation of transgenic products. The Japanese have expressed equally strong doubts about the safety and appeal of genetically engineered foods, and the government there is also planning to have such foods labeled. Foreign resistance to genetically modified foods has caused trade conflicts with the United States, where most such foods are produced.

In an attempt to soothe fears in Europe and, to a growing extent, in the United States as well, U.S. Secretary of Agriculture Dan Glickman issued a statement on July 13, 1999, in which he announced that the government would conduct long-term studies on the ecological and health effects of genetically altered farm products for the first time. Glickman also asked for an independent, scientific review of the department's system for approving genetically engineered crops. He said he planned to urge the American biotechnology industry to voluntarily label products containing genetically altered material. At the same time, he also noted that the Clinton Administration would use all available legal means, including tariffs on European-made goods, to compel Europe to accept American farm products, including those that may contain bioengineered material. Groups expressed concern about genetically modified foods at both the World Trade Organization talks in Seattle and an FDA hearing in Oakland, California, in December 1999.

Transgenic farm animals have existed since 1985, although the techniques for producing them are still experimental and often fail. Some have been altered to increase their value as food (to produce leaner meat, for example), but many are intended for the branch of the biotechnology industry often called "pharming." They have been given genes to induce production of human hormones, drugs, or other medically useful substances in milk (cattle), eggs (chickens), or even urine (mice). Producing such substances in animals costs less than 10 percent as much as making them in laboratories or factories, and the substances often can be made more quickly as well.

Pharming is not the only use of genetically altered animals. Mice, for instance, have been given human genes associated with cancer or other diseases to make them more useful in experiments aimed at understanding the diseases or in testing drugs. Pigs have been given human genes in the hope that the animals may become compatible donors for organ transplants.

The technology for creating genetically altered animals received a large boost in February 1997, when Ian Wilmut and his coworkers at Scotland's Roslin Institute announced that they had cloned a lamb from an udder cell of a six-year-old Finn Dorset ewe. They named the lamb Dolly, after country-western singer/actress Dolly Parton. The amazing thing about Dolly, from a scientific point of view, was not that she was a clone—a sort of delayed twin of her "mother," with exactly the same genetic makeup—but that she had been made from a mature adult somatic (body) cell, something many scientists had thought could not be done with mammals.

Scientists had first attempted to clone amphibians in 1952. In 1986 they developed a new cloning technique called nuclear transfer, in which a single embryo cell is fused by electricity with an unfertilized egg from which the nucleus has been removed. The technique was normally used with embryo cells, but Wilmut and his team found they could use it on a mature cell if they first deprived the cell of nutrients for five days. This starvation put the cell into a resting state, in which it did not divide and most of its genes were turned off. After the cell was fused with an enucleated egg, the cytoplasm of the egg cell somehow reprogrammed the adult cell's nucleus. The combined cell eventually produced a whole animal.

Issues in Biotechnology and Genetic Engineering

Wilmut's chief aim, and that of many scientists who followed him, was to create a more efficient way to produce herds of identical, genetically altered "pharm" animals. Cloning is potentially better than breeding for producing such animals because with cloning a researcher can be sure that a particular gene or trait existing in the original animal will be preserved, whereas it may or may not be in breeding. Some scientists have also suggested cloning as a way to preserve endangered species or even, perhaps, revive extinct ones, such as mammoths.

Within months, Dolly was followed by cloned cattle and other animals, including the first cloned transgenic farm animals (which were also sheep). Most were created with variations of Wilmut's technique. The cloning technique, like techniques for inserting foreign genes into animal germ (reproductive) cells, is still in the early stages of development and very inefficient, however. Dolly, for instance, was the only success out of 277 tries. Some later-cloned animals have died mysteriously soon after birth. Many scientists suspect that embryonic cells will remain better candidates for cloning than will mature cells. This suspicion comes partly from a follow-up study published in May 1999 that revealed that, at least by one measure, the age of Dolly's cells matched that of her mother rather than her own.

So far, transgenic animals have produced much less controversy than transgenic plants, perhaps because these animals have been made in such small numbers. Most complaints have focused on health risks to the animals themselves. For instance, animal rights groups have pointed out that a "fast-growing" pig containing growth hormone genes from cows (created by the USDA in the 1980s), was deformed, crippled by arthritis at a young age, and unable to keep itself warm because its body contained so little fat. Such groups also question the ethics of treating animals as mere factories or machines for the production of drugs.

The greatest safety controversy on the animal side of agricultural biotechnology has centered, not on animals that contain altered genes, but on a genetically engineered substance given to normal animals. Beginning in 1985, Monsanto developed a process for making genetically altered bacteria produce bovine growth hormone (BGH, or bovine somatotropin, or BST). The recombinant hormone (rBGH), trade-named Posilac, is sometimes given to dairy cattle to increase their milk production. It won approval from the FDA in 1993, thereby becoming the first genetically engineered animal hormone approved for use in the United States. No other industrialized country has approved its use.

The FDA has ruled that rBGH is not an additive and therefore does not have to be listed on the label of milk taken from treated cows. A number of groups in the United States and elsewhere have questioned that decision, however. They point out that the hormone is associated with an increased

17

risk of mastitis (inflammation of the udder) and other health problems in treated cows. To prevent or treat the mastitis, dairy farmers give cows antibiotics, some of which can remain in their milk and be drunk by humans. Constant exposure to low doses of antibiotics encourages bacteria that infect humans to become resistant to the drugs. Some critics also worry that insulinlike growth factor 1 (IGF-1), a hormone related to BGH, may be absorbed from the milk of rBGH-treated cows. This substance may increase the risk of breast and prostate cancer.

Fears of rBGH have led to some legal battles. The state of Vermont, suspicious of rBGH from the time the hormone was first tested there in the mid-1980s, passed a law in 1994 requiring that stores label all products that might come from hormone-treated cows. Six industry groups challenged the law in a federal suit, claiming that it interfered with their free speech and with interstate commerce. A U.S. District court in Vermont upheld the law in 1995, but a year later the Court of Appeals for the Second Circuit overruled it. Vermont now has a voluntary labeling law that allows farmers to state that their milk does not come from rBGH-treated cows.

PATENTING LIFE

Herbert Boyer and Stanley Cohen did not think of patenting their revolutionary gene-splicing technique until their universities insisted that they do so. When they finally applied for the patent in 1974, they donated all proceeds from it to the universities. This altruistic picture did not remain the norm for long, however. Profit motives entered the scene in 1976, when Robert Swanson, a young venture capitalist, persuaded Boyer to join him in founding Genentech (GENetic ENgineering TECHnology), the first biotechnology company based on genetic engineering.

Like many similar businesses that sprang up in the following years, Genentech took advantage of the remarkable reproductive powers of bacteria, the first genetically engineered organisms, and the fact that techniques for growing bacteria in factories already existed. Doubling in number every 20 minutes, bacteria could produce billions of duplicates of themselves—including any genes that had been inserted into them—in just a day or so. All of these clones could make the proteins for which the inserted genes coded.

Swanson and Boyer planned to use genetically altered bacteria to make human hormones or other medically useful substances that had been in short supply. They started with insulin, the sugar-controlling hormone that many diabetics must take daily. Insulin from cattle or pigs was easily available, but the animal forms of the hormone differ slightly from the human version, and about 5 percent of diabetics are allergic to them. The product of bacteria containing the human insulin gene, by contrast, was identical to the human hormone and therefore caused no allergic reaction.

When Genentech offered its stock to the public for the first time in October 1980, it had produced genetically engineered human insulin experimentally (starting in 1978) but had not yet won federal permission to sell it. Nonetheless, belief in the company's future was so strong that the price of its stock jumped from $35 to $89 a share in the first few minutes of trading. One newspaper called Genentech stock "the most spectacular new stock offering in at least a decade."[7] By the time Genentech finally won FDA approval to sell its genetically engineered insulin in 1982, companies that had followed its lead were using modified bacteria to produce some 48 different hormones or other substances from the human body, as well as several drugs and vaccines. The business side of genetic engineering was well under way.

Where money goes, lawyers and lawsuits soon follow, and biotechnology has been no exception. Most of the suits and other legal disputes in the industry have concerned patents and other ways to protect intellectual property. Genetic engineering has broken new ground in the patent field by raising the question of when—or, indeed, whether—living things, body parts, cells, genes, or DNA sequences should be patented. Is it possible or right to "own" a piece of life?

Patents are a time-honored way of rewarding and encouraging inventiveness by giving an inventor the exclusive rights to an invention's use and sale. The doges of Venice and the rulers of England, among others, traditionally granted monopolies, or rights of exclusive sale, to inventors. The term *patent* came from the "letters patent," or open letters addressed to the general public, that kings used to proclaim these monopolies.

Article 1, section 8 of the U.S. Constitution grants Congress the right to "promote the progress of science and useful arts, by securing for limited times to authors and inventors the exclusive right to their respective writings and discoveries." Patents are issued by the Patent and Trademark Office (PTO), which is part of the Department of Commerce. In 1793, Congress defined a patentable invention as "any new and useful art, machine, manufacture or composition of matter." The Patent Act of 1952 changed the word "art" to "process" and added the criterion that, as well as being new and useful, an invention must not be obvious "to a person of ordinary skill in the art."[8]

Traditionally, products of nature, including living things, were not held to be patentable because they were not inventions, or objects created by human ingenuity. An exception of sorts was made for plant varieties, which were protected in the United States by the Plant Patent Act (1930) and the Plant Variety Protection Act (1970). These laws allowed breeders who had developed new varieties to block others from reproducing them.

Biotechnology and Genetic Engineering

Genetic engineering brought a major change to the United States patent system in 1980, when the Supreme Court ruled by a 5 to 4 vote that a patent could be issued to Ananda Chakrabarty, a scientist working for General Electric Corporation, for a genetically altered bacterium that digested petroleum and could be used to help clean up oil spills. In his majority opinion, Chief Justice Warren Burger cited evidence that Congress had intended the 1952 patent law to apply to "anything under the sun that is made by man"— including living things, if humans had altered them.[9] Chakrabarty's patent was the first patent issued for a living organism.

The *Diamond v. Chakrabarty* ruling was soon extended to more complex organisms. Plants, seeds, and plant tissue cultures became patentable in October 1985, when Molecular Genetics, Inc., obtained a patent on a type of genetically engineered corn. A year and a half later, in April 1987, the PTO ruled that genetically engineered animals (except humans), as well as human genes, cells, and organs, could also be patented. The first genetically engineered animal to be patented was the Harvard Oncomouse, a type of mouse designed to be a test animal in cancer research. It was patented in April 1988 by Harvard University. The PTO received 1,502 patent applications for transgenic animals in the decade that followed and approved more than 90. Most were genetically altered mice intended for medical research, but the list ranged from worms to sheep.

Not everyone has agreed that patenting living things is ethical. In May 1995, for example, antibiotech gadfly Jeremy Rifkin organized a coalition of about 200 religious leaders representing 80 faiths who gathered in Washington, D.C., to present a "Joint Appeal Against Human and Animal Patenting" and called for a moratorium on this activity. "We believe that humans and animals are creations of God, not humans, and as such should not be patented as human inventions," their petition stated.[10]

Other criticisms of the biotechnology industry's stress on patents, including those on living things, have come from scientists. They point out that, in the United States at least, an invention cannot be patented if details of it have been published more than a year before the patent application is filed. Researchers thus often withhold data until applications can be submitted, impeding the free flow of information that has been traditional in science. This complaint is ironic considering that patents—which require a description of an invention detailed enough to allow anyone skilled in its field to make the device—were intended as a way of increasing access to information about technology. Furthermore, as more and more scientists, including those whose main work is done at universities, become involved in biotechnology ventures, the urge to patent may make conflict of interest affect the objectivity of their work. At the very least, such conflicts affect others' perception of the scientists' objectivity. For example, the objectivity of a scientist's study of the effectiveness of a drug might be questioned if the scientist was part owner of a patent on the drug or had stock in a company that manufactured it.

Patents also create monopolies that can reduce the supply of products, increase their price, and limit competition. This is potentially a particular problem in agricultural biotechnology, where giant companies such as Monsanto and Du Pont often control both genetically engineered crop seeds and the chemicals on which they depend. Some farmers, especially in the Third World, may not be able to afford these products or to compete against those who can.

This situation becomes even more politically charged when the patents are for altered—or even, sometimes, unaltered—forms of living things that originally came from the poverty-stricken tropics. Critics such as the Rural Advancement Foundation International (RAFI) claim that large companies perform acts of "biopiracy," raiding the developing world—the source of most of the planet's genetic diversity and home to most of its undiscovered species—for specimens or genetic samples that may prove useful. These companies, the critical groups say, provide little or nothing in return to the land that nurtured the plants or the native farmers and healers who used or developed them.

Protests have arisen over exceptionally broad patents, beginning with the "process" and "product" patents granted to Boyer and Cohen, which seemed to cover all of recombinant DNA technology. A later dispute of this kind related to patents granted in the early 1990s to Agracetus, a subsidiary of chemical giant W. R. Grace, for all genetically engineered cotton and soybeans, regardless of the method by which they were produced. "It was as if the inventor of the assembly line had won property rights to all mass-produced goods," complained Jerry Caulder, CEO of a rival company, Mycogen.[11] A related subject of dispute is so-called reach-through rights, which give the owner of a patented technology royalties on all commercial products created by that technology, regardless of who develops them. Companies often try to obtain reach-through rights from licensees as a condition of licensing their technologies. Complaints from other companies or, sometimes, government bodies such as NIH have caused the PTO to reverse its decisions on some broad patents, such as one of those granted to Agracetus, and have forced some patent-holding companies to moderate their demands for reach-through rights.

In answer to these complaints, biotechnology industry spokespeople insist that patents on genetically altered living things and the processes that produce them are both ethical and necessary. Patents, they note, traditionally have been held to benefit society by encouraging invention and giving inventors an inducement to risk the time and money necessary to bring inventions to the marketplace. Patents also forced inventors to make public the details of their technology, thus (once the patent expired) encouraging business expansion and competition. Patents, supporters claim, are necessary to give modern biotechnology investors an incentive to support the research and testing needed to develop, say, a new drug. Such a development process can take 7 to 10 years and cost up to $500 million. Biotech boosters claim that society ultimately benefits from industry patents because patented research can produce such things as increases in world food supply and life-saving medicines.

As for "biopiracy"—or "bioprospecting," as its supporters prefer to call it—both sides of the debate cite as an example an agreement made in September 1991 between the National Biodiversity Institute (INBio) in Costa Rica and American drug giant Merck & Co. INBio promised to provide samples of Costa Rican wild plants, microbes, and insects—indeed, what foes say is the entire genetic wealth of the country—to Merck to screen for possibly useful drugs. In return, Merck offered to give INBio a research and sampling budget of $1.135 million, royalties from any commercial products that result from the samples, and training and assistance to help Costa Rica set up its own drug screening program. INBio promised to give 10 percent of its initial payment and 50 percent of its royalties to a fund used to conserve the country's national parks. Supporters say that this sort of agreement provides excellent incentives for countries to preserve their biological diversity, whereas opponents see it as selling a country's genetic birthright. Several international agreements, such as the Convention on Biological Diversity and Intellectual Property Rights, signed by 157 nations (not including the United States) at the Earth Summit in Rio de Janeiro in June 1992, have included sections that attempt to protect countries' biological and genetic resources.

As with labeling of genetically engineered foods, differences in public feeling and law between the United States and other parts of the world have caused conflicts in regard to patenting living things. For instance, Europe does not allow patenting of genetically modified animals unless "substantial medical benefit" is derived from doing so, although altered plants can be patented. In international talks such as those on the General Agreement on Tariffs and Trade (GATT) in 1994, the United States has pressured other nations to strengthen their laws on protection of intellectual property in biotechnology.

Types of human cells and natural human body chemicals have been patented in the United States by companies that developed processes for isolating them, even if the cells or chemicals themselves were not changed. Similarly, an amendment to a draft directive on intellectual property rights in gene technology approved by the European Parliament in July 1997 forbids patenting the human body, its elements, or products but states that "an element isolated from the human body or otherwise, produced by means of a technical process shall be patentable even if the structure of that element is identical to that of a natural element."[12] When a line of laboratory cells can be traced to the body of a particular individual, however, disputes have arisen about who owns the rights to it.

The most widely discussed case of this kind in the United States began in 1984, when Seattle businessman John Moore sued his doctor, the University of California, and several drug companies. Moore's suit claimed that the drug companies had made and patented a profitable laboratory cell line from his spleen, which had been surgically removed as a cancer treatment in 1976. Moore claimed that, although he had consented to the operation, he had not been told about the commercial use of his spleen cells, which he said had been planned even before the surgery. He maintained that the cells were still his property even after they had been removed from his body and that they should not have been used without his permission.

A lower court supported Moore, but in 1990 the California Supreme Court ruled against him, stating that patients do not have property rights to their tissues. The court's majority opinion said that granting such rights would "destroy the economic incentive to conduct important medical research."[13] Not all states have followed California's policy, however. In July 1997, for instance, Oregon passed a law granting ownership of tissue and all information derived from it to the person from whom the tissue came.

Even greater outcries have arisen when the person whose cells were patented has been a member of the developing world. The first such case occurred in 1993, when RAFI got word that the NIH had applied for a patent on a cell line derived from a Guaymi (Panamanian Indian) woman. The NIH hoped that the cell line would prove useful in combating HTLV-1, a virus that causes leukemia in humans, or even possibly AIDS, which is caused by a related virus. Protests from the Guaymi General Congress in Panama and others forced the NIH to drop the patent request, however. A similar uproar occurred about a year later when the NIH tried to patent another cell line made from the blood of a man of the Hagahai people in Papua New Guinea. The NIH initially this time stuck to its guns and was awarded the patent in March 1995, but continued objections made it drop the claim in 1996.

Concerns about "biocolonialism" have also dogged the Human Genome Diversity Project, which was established in 1992 with private funding and headed by Stanford University geneticist Luca Cavalli-Sforza. The project aims to collect DNA (in the form of hair, saliva, and blood samples) from about 500 of the world's most endangered indigenous peoples and preserve it in the form of cell lines. Cavalli-Sforza and the project's supporters claim that it will provide information about, for instance, past human migrations and differences in susceptibility to diseases that will benefit the whole world. However, Cultural Survival, an indigenous peoples' support group based in Canada and Massachusetts, calls it the "vampire project" and claims that it is merely an attempt by large multinational companies to exploit the information in indigenous people's DNA for profit. A similar project, launched in 1999 by a coalition of 10 multinational drug companies, a London-based charity, and five genetics laboratories, will no doubt be met with similar criticisms.

In the 1990s the cutting edge of biotechnology patenting moved from organisms and cells to pieces of the genetic code itself. Aided by such inventions as the automatic gene sequencer and the polymerase chain reaction (PCR), which allows tiny pieces of DNA to be rapidly copied many times for analysis, scientists in the late 1980s began to work out the base sequences of the genomes, or complete collections of genes, of living things. This activity led to the 1990 launching of the Human Genome Project (HGP), which intends to sequence all 80,000 to 140,000 human genes, comprising some 3 billion base pairs. The $3 billion project, on which dozens of laboratories worldwide are cooperating, is expected to be complete by 2002. The choice of this date, a three-year advance on the original deadline of 2005, was prompted by competition from several private ventures, which in 1998 claimed that they would achieve the genome project's goal sooner and more cheaply than the government project. In January 2000, one of those companies, Celera Genomics, announced that it already had at least 90% of human genes in its database and expects to complete its sequencing of the human genome by summer 2000.

The United States PTO has permitted the patenting of genes or DNA sequences, including human ones, since 1987, but these genes have normally been required to fulfill the standard patent requirements of novelty, usefulness, and unobviousness. A considerable dispute erupted in 1991, when the NIH applied for a patent on some 2,700 human DNA sequences called expressed sequence tags (ESTs)—the active parts of a cell's DNA. Craig Venter, a scientist then working for NIH, had isolated the ESTs from an existing library of brain genes with a computer-assisted technique he had invented. Opposition to this patent application came from a variety of sources, ranging from biotechnology trade organizations to DNA pioneer James Watson, who at the time headed the HGP. The PTO eventually rejected the NIH application on the grounds that the genetic fragments met none of the patent requirements: They were not novel because they came from an existing DNA library, their function (utility) was not known, and they were "obvious" to anyone who could use Venter's technique.

In spite of this rejection and of early opposition to wholesale patenting of DNA sequences from such groups as the American Society of Human

Genetics, which feared that patenting would tie up useful research information and put an undue emphasis on sequencing genes for profit, mass gene patenting has continued. The PTO received 500,000 requests for patents on nucleic acid sequences in 1996 alone, and more than 1,800 genes had been patented by 1998. Three ESTs had also been patented, and more were expected to follow. The applicant for a patent on a gene or DNA sequence must specify the sequence precisely and show that it is potentially useful, but the claim for usefulness can be very vague, such as saying that the gene can be used as a probe to find other genes. Patenting of genes is also permitted in Europe, though the draft directive on intellectual property rights approved by the European Parliament in July 1997 bans patents on human genes.

DNA "FINGERPRINTING"

DNA "fingerprinting" has been used to identify lost children and straying fathers, reveal the genetic makeup of endangered species, and trace the origins of poached elephant tusks. It proved that bones found in Siberia belonged to the last Russian czar and disproved the claims of a woman who had insisted for decades that she was his surviving daughter. More important to most people, it has convicted—or exonerated—hundreds of people accused of rape and murder. DNA "fingerprinting" is also the most direct application of biotechnology to the courtroom.

DNA profiling or identification testing, as this group of techniques is more frequently known, was invented by Alec Jeffreys, a geneticist at the University of Leicester in Britain, in the early 1980s. In 1985 Jeffreys published a paper describing it in the prestigious British scientific journal *Nature*. He based his test on the fact that certain regions in human DNA consist of short base sequences repeated over and over. These regions are not the protein-coding part of genes (exons) but rather so-called junk DNA (introns), the biological purpose of which (if any) is still unknown. Jeffreys first used regions called minisatellites, but identity testing soon focused instead on other stretches of DNA called variable number of tandem repeats (VNTRs). Unlike most genes, which exist in only a few different forms, or alleles, VNTRs vary considerably from person to person, even within the same family, in the number of repeated sequences they possess. The chance that two people, other than identical twins (who have exactly the same genetic makeup), would have the same numbers of repeats at, say, five different spots is vanishingly small. Comparison of DNA from a blob of semen or a bloodstain left at a crime scene with the blood of a suspect, therefore, should be able to show with high accuracy whether the crime evidence came from the suspect.

Biotechnology and Genetic Engineering

Jeffreys recommended that DNA be analyzed by one of two methods that were already becoming vital to molecular biology. One method, restriction fragment length polymorphism (RFLP, pronounced "riflip") analysis, was starting to be used to locate genes that caused inherited diseases. Developed in 1978, it uses restriction enzymes (those same "molecular scissors" that had been so helpful to Herbert Boyer and Stanley Cohen when they began splicing genes) to cut the DNA being tested into small pieces. Because some genes have several alleles that differ in the number of bases they contain, segments of DNA from different people containing these genes will have different lengths when cut with the same restriction enzyme.

The RFLPs—specifically, in the case of DNA profiling, the VNTRs—are analyzed by a standard laboratory technique called electrophoresis. The DNA fragments produced by the restriction enzymes are placed on a gel and zapped by an electric current. The electricity makes the fragments move through the gel at speeds that depend on their molecular weight, with the lighter ones moving more quickly and the heavier ones trailing in the rear. The fragments, now spread out in a series of bands, are blotted onto a nylon membrane, and radioactively labeled probes, consisting of DNA with known variations at the tested spots, are applied. If the test sample contains base sequences that match those of the probes, the probes attach themselves to those sequences. The membrane is then exposed to X-ray film, and radioactivity from the bound probes makes spots on the film. The resulting picture, called an autoradiograph, is a series of light and dark bands, much like a supermarket bar code. DNA profiling by the RFLP method usually examines four to seven spots, or loci, containing different VNTRs. Technicians compare autoradiographs from crime scene and suspect DNA through the human eye, a computer, or both, to detect pattern matches.

When done properly, RFLP analysis is extremely accurate. Unfortunately, it is also difficult and time consuming, often taking two or three months, and it requires a relatively large test sample in good condition. DNA at a crime scene, by contrast, often exists in minute amounts that have been contaminated with bacteria or exposed to damaging environmental factors such as sunlight. To analyze this kind of sample, testers use a second, somewhat less accurate but faster technique that centers on the polymerase chain reaction, or PCR.

American molecular biologist Kary Mullis discovered PCR in 1983. Using an enzyme called polymerase, the process acts as a sort of genetic duplicating machine, doubling the test DNA each time it is repeated. It can multiply a sample a trillion times in just a few hours. Portions of the beefed-up sample are then applied to 8 or 10 spots on reagent strips, each of which contains a different segment of known DNA. If the replicated DNA matches the known DNA on a spot, a blue dot appears there. A technician then compares the pattern of spots formed by the crime scene sample to the pattern formed by a suspect's DNA. The test's chief drawback is that its accuracy can be very easily thrown off by contamination of the sample with a tiny bit of DNA from, say, another piece of crime scene evidence, blood given by a suspect, or a different sample previously tested in the same equipment.

Jeffreys' DNA profiling technique was first used to verify family relationships in immigration and paternity disputes. It entered the criminal court in 1987, when, faced with the rape and murder of two teenaged girls, police in Leicestershire took the unprecedented step of asking all men between the ages 13 and 30 years in three villages near the crime scenes—about 5,000 people—to give blood samples for comparison with semen found on the bodies of the girls. At the officers' request, Jeffreys agreed to analyze all the specimens that could not be quickly eliminated on the basis of blood type, some 40 percent of the total.

While Jeffreys was testing, police received word of a conversation in a local pub in which a man had bragged about having given a blood sample in a coworker's stead. They questioned the man, and he identified the reluctant coworker as a 27-year-old baker named Colin Pitchfork. Faced with evidence of his falsification, Pitchfork confessed to the crimes, and when his blood was finally tested, its DNA matched that in the semen samples. Pitchfork went to trial in September 1987, was found guilty, and was sentenced to life in prison, thus becoming the first criminal convicted by DNA evidence. DNA testing showed its other side—its ability to clear innocent people—in the same case by demonstrating that 17-year-old Rodney Buckland, who had confessed to one of the murders, could not have committed it because the DNA in his blood did not match that in the semen samples.

The first American trial in which DNA testing convicted a criminal was also a rape case. In 1987, a 24-year-old warehouse worker named Tommie Lee Andrews was accused of stalking several women in Orlando, Florida, breaking into their homes at night when he knew they would be alone, and then beating and raping them. Prosecutor Tim Berry sent samples of semen from one of the rape victims and Andrews's blood to Lifecodes, one of the few American laboratories that performed DNA identification testingat the time. The samples matched.

The judge agreed to admit the test results as evidence in Andrews's trial, which began on October 20, 1987, after an expert witness claimed that DNA testing, although new to forensics, was a standard procedure in genetics laboratories. During the trial itself, however, the judge refused to allow an expert called by Berry to testify about the statistical probability of Andrews's DNA pattern matching that of the rape specimen by chance, which was estimated at 1 in 10 billion. The jury could not agree on a verdict, and the judge declared a mistrial. Andrews went on trial for one of the other rapes in November, however, and this time the statistical testimony was allowed and Andrews was found guilty on November 6. He was also found guilty of the first rape charge in a retrial in February 1988. His sentences for the crimes added up to one hundred years in prison.

Few people questioned DNA testing at first. Police and prosecutors regarded it as the greatest aid to criminal identification since the introduction of fingerprinting a century before. Expert testimony that the odds of a suspect's DNA matching that in a crime scene sample by chance were, say, one in a trillion awed judges and juries alike. As one juror in an early DNA case said, "You can't argue with science."[14]

Within a year or two, however, people did begin to argue with science. No one, then or since, disagreed with the basic principles behind DNA testing, but lawyers and scientists questioned both the accuracy with which the tests were carried out and evaluated and the methods used to derive the statistics that sounded so impressive in a courtroom. A major case in which such questioning occurred was *New York v. Castro* (1989), in which the expert witnesses for the prosecution and defense took the unprecedented step of meeting outside the courtroom and jointly concluding that the DNA evidence in the case was worthless. Another case was *United States v. Yee* (1991), in which disputes about the scientific validity of some aspects of the testing shaded over into personal attacks on the scientists who testified.

From DNA profiling's first uses to the present day, the accuracy with which the tests are carried out has been questioned. Sometimes accuracy is fatally compromised by the mishandling of samples before they even reach the testing laboratories, as the defense suggested so successfully in the famous O. J. Simpson murder trial in 1995. Testing laboratories themselves vary in quality and, until 1994, despite repeated requests from scientists in the field, there was no federal program for licensing or giving proficiency tests to laboratories that performed DNA analysis. In that year, Congress passed the DNA Identification Act, which gave responsibility for training, funding, and testing DNA profiling laboratories to the Federal Bureau of Investigation (FBI). Critics have questioned the wisdom of this, both because the FBI's guidelines, established in 1991, stress self-testing of laboratories rather than proficiency testing by an outside agency and because the FBI's own DNA testing laboratory has been shown to have, as Deputy Attorney General Jamie Gorelick admitted in January 1997, "a serious set of problems" in the accuracy of its work.[15]

Interpretation of the often-ambiguous results of DNA tests has also raised questions, especially when technicians compare samples by eye rather than by computer. Around 1990, for instance, laboratories using the FBI's procedures made two tests of samples from the same group of FBI agents 14 months apart and failed to find matches between two samples from the same person in 12 percent of the cases.

Problems of interpretation can arise from the fact that the bands on autoradiographs may vary somewhat in size, even when they are considered to match. Confusion can also be caused by "bandshifting," in which one sample moves through the electrophoresis gel at a different rate of speed than the other and its entire pattern of bands is therefore displaced up or down. Testers can correct for bandshifting by including a locus that does not vary and thus can be guaranteed to match between samples. Measuring the displacement at that spot can show the displacement of the overall sample. Comparison of DNA profiles remains difficult, however.

The most serious disputes about the scientific side of DNA testing have concerned the statistics used in court to show the probability of an accidental match between suspect and crime scene samples. To begin with, as critics have pointed out, the frequently used term DNA "fingerprinting" (coined and trademarked by Jeffreys) is somewhat misleading. Each person's fingerprint is unique, and so (unless the person has an identical twin) is each person's DNA. DNA identification tests, however, do not examine a person's whole genome, but only tiny fragments of it. The chance that someone else will have the same sequence in those fragments is usually very small, but not zero, especially if two suspects belong to the same family or small subset of an ethnic group. Thus, although lack of a match in a DNA test can conclusively prove innocence (assuming that the test was properly carried out), a match does not conclusively prove guilt.

Confusion about DNA statistics can be caused by what has been called "the prosecutor's fallacy." There are two questions to be asked of any DNA evidence: First, what is the probability that an individual's DNA will match that in the crime scene material, given that the person is innocent? Second, What is the probability that an individual is innocent, given that there is a match? The fallacy is giving the answer to the first question in response to the second. The statistics usually given in court state the probability of finding two matching DNA profiles by chance in a population, and they are not always as impressive as they sound; if the probability of two profiles matching by chance is 1 in 20,000, for instance, then 250 such profiles theoretically could still be found in a city of 5,000,000 people.

The liveliest debate—or, as John Hicks, then head of the FBI crime lab more bluntly put it, war—about the accuracy of the statistics quoted by DNA experts, however, was more arcane than that. Opposing camps of population geneticists fought this war in court and on the pages of scientific journals between 1990 and 1992. Before it was over, personal feelings, scientific reputations, and judicial faith in DNA testing had all been severely battered.

Biotechnology and Genetic Engineering

The odds against two DNA profiles matching by chance are traditionally figured by the so-called multiplication rule: that is, the frequencies of all the tested alleles in a broad target population such as "Caucasians" or "Blacks" are multiplied together. If a particular allele at each of four loci occurs in 1 out of 100 people in the target population, for instance, the probability of two people having those same alleles at all four loci is 1 in 100 x 100 x 100 x 100, or 1 in 100,000,000. The multiplication rule is accurate only if each allele is inherited independently of all the others. The assumption that this is true, in turn, is based on the assumption that people within the target group intermarry randomly.

That underlying assumption was what some population geneticists, most notably Richard Lewontin of Harvard University and Daniel Hartl, then of Washington University in St. Louis, questioned. Lewontin and Hartl pointed out, first in testimony in the *United States v. Yee* pretrial hearing in the summer of 1990 and then in an article in the December 20, 1991, issue of *Science* magazine, that a large group such as "Caucasians" is really made up of many ethnic subgroups with different origins. People tend to marry within their own subgroup, the geneticists said, not randomly within the larger population. As they put it, "Americans tend to marry the girl or boy next door."[16] That fact could throw off the statistics produced by the multiplication rule by two or more orders of magnitude, they claimed.

Lewontin and Hartl recommended that DNA data from a variety of subpopulations be gathered to find out whether these groups had different frequencies of alleles at the loci most often used in DNA testing, a process that could take 10 to 15 years. In the meantime, they said, experts testifying in court should use calculation methods that produced more accurate, even if less impressive, figures than the multiplication rule. Opponents of their point of view, including Ranajit Chakraborty of the University of Texas and Kenneth Kidd of Yale, who wrote a rebuttal article printed in the same issue of *Science*, replied that such caution was not necessary. The existence of ethnic subgroups might well affect marriage patterns, but these geneticists claimed that such variations did not significantly affect the independent inheritance of the alleles and therefore did not invalidate the multiplication rule.

This seemingly technical argument became very personal in the pretrial hearing to determine the admissibility of the DNA evidence in the *Yee* case. Lewontin, Hartl, Kidd, and Thomas Caskey of the Baylor College of Medicine in Houston (who was on the Chakraborty-Kidd side of the debate) all testified, and all, either during the hearing or afterward, found not only their scientific points of view but also their individual integrity attacked by the opposing lawyers. Accusations focused on such things as the high fees paid for some of the scientists' testimony and links between others and the

FBI or private testing labs. Such adversarial behavior is common in court-rooms but not usual in scientific debate, and some of the scientists were shocked by it. For instance, Hartl, who had not testified in court before, found the experience "emotionally draining and harrowing."[17]

The *Science* articles grew out of the *Yee* debate and begot their own round of accusations and counter-accusations. Lewontin and Hartl claimed that the Justice Department tried to intimidate them into withdrawing their article or talk *Science* editor Daniel Koshland into not publishing it. Koshland said the agency had never even contacted him, let alone intimidated him, and claimed that he asked Chakraborty and Kidd to write a rebuttal article simply to present both sides of the debate.

This widely publicized scientific fracas threatened to have a major effect on the way courts viewed DNA evidence. Judges traditionally used the so-called Frye rule (named after a 1923 case, *Frye v. United States*) in deciding whether to accept evidence based on new technologies. The judge in the Frye case had stated that

> *while courts will go a long way in admitting expert testimony deduced from a well-recognized scientific principle or discovery, the thing from which the deduction is made must be sufficiently established to have gained general acceptance in the particular field in which it belongs.*[18]

If evidence from a new technology is expected to play a key role in a trial, a judge usually holds a pretrial "Frye hearing" to determine the admissibility of that evidence. The scientific debate in the *Yee* case took place in a Frye hearing, for instance. (It should be noted that the Frye rule has always had its critics, who have complained that the "general acceptance" criterion is vague and places more emphasis on consensus than on scientific validation. The Supreme Court has stated that judges do not have to use this rule in determining the admissibility of scientific evidence.) Most judges had been allowing DNA evidence in their courts, but the population genetics dispute was bound to make them wonder whether there really was a "scientific consensus" about this technology.

Attempting both to soothe ruffled professional feelings and to preserve DNA testing as a tool for law enforcement, a committee from the National Research Council (NRC) of the National Academy of Sciences hurried to finish a report on the subject that they had been working on since the *Castro* debacle in 1989. They issued the report in May 1992. To deal with the issue of testing accuracy, they recommended that the Department of Health and Human Services, in consultation with the Justice Department, oversee mandatory proficiency testing and accreditation of forensic DNA laboratories. An expert committee appointed by NIH or the National Institute of

Standards and Technology could evaluate new forms of testing, they said, thus doing away with the necessity for lengthy and contentious Frye hearings. Finally, to settle the population genetics issue, the committee suggested a compromise "ceiling principle" calculation method that could be used until accurate figures for frequencies of alleles in different subpopulations were obtained.

The NRC committee believed that the ceiling principle would provide a "conservative" answer to the probability question, overstating the frequency of most alleles and thus favoring defendants. "I don't think anyone will fight it," said committee member Eric Lander, a molecular geneticist at the Whitehead Institute in Cambridge, Massachusetts.[19] Lander was wrong. A number of population geneticists and mathematicians in fact lambasted the principle, saying that it was arbitrary and favored defendants too much. Congress, meanwhile, ignored the committee's recommendations about proficiency testing and evaluation of new tests.

The NRC issued a second report in 1996 in which they abandoned the ceiling principle, stating that accurate enough information on allele frequency in a variety of ethnic subgroups was now available to allow calculations to be done on the basis of these subgroups. If the ethnic group of the perpetrator was unknown, separate calculations for each potential ethnic group should be made and the judge should decide which to use, the committee said. David Kaye, a law professor at Arizona State University who was a member of the committee, said, "I believe this report will carry a lot of weight with judges because it is the voice of a consensus of scientific opinion," thus fulfilling the requirement of the Frye rule.[20]

This time the optimistic prediction seemingly was right; the 1996 report produced none of the strong objections that greeted the 1992 one. By then, most researchers had come to agree that as long as enough different loci (at least eight) are tested, "the statistical issues recede into the background," as Daniel Hartl said in 1994.[21] The chief negative comments about the second NRC report centered, not on population genetics, but on its elimination of the first report's demand for standardized, blind proficiency testing of DNA laboratories. The 1996 report claimed that such testing would impose "formidable" logistical demands, but the outspoken Richard Lewontin accused the committee of having been "bought" by the Department of Justice, which did not feel that outside testing was necessary.

There is still no national system of outside oversight and proficiency testing for forensic DNA laboratories, and the quality of sample handling and testing is still sometimes an issue in court cases. However, many police departments, embarrassed by revelations such as those in the Simpson trial, have tried to improve their procedures. For instance, officers may use different disposable tweezers and gloves for picking up each sample, reducing the possibility of contamination.

Techniques of testing have also improved. The preferred form of test today, developed by Thomas Caskey in the mid 1990s, focuses on so-called short tandem repeats (STRs) and has some features of both the VNTR/RFLP and the PCR methods. It can check 13 different loci, even in a sample as small as that obtained from a bit of saliva or a single hair root. Like PCR, it is very susceptible to contamination, but devices now under development may reduce the risk of error by completely automating the test.

The time seems to be coming, if it has not already arrived, when properly carried out DNA profiling will provide the (for all practical purposes) unique "fingerprint" that Alec Jeffreys claimed. Although DNA evidence is used in only a small percentage of criminal trials, such evidence had, by the late 1990s, helped to convict hundreds of criminals as well as to clear dozens of other people who had been mistakenly accused or convicted. "So long as you have a good sample and a competent lab following appropriate procedures," Northwestern University criminal law professor Ronald Allen said in 1998, "DNA evidence is devastating."[22]

Today the chief subject of debate related to DNA testing is the preservation of samples or test results in large databases, a growing practice that concerns many civil libertarians. All 50 states now have laws requiring collection and storage of DNA samples from convicted murderers, rapists, and child molesters, and in October 1998 the FBI set up the Combined DNA Index System (CODIS), a database intended to combine information from all the state databases so that police can match samples at crime scenes with DNA profiles of convicted felons anywhere in the country. Since 1995, Britain has had a similar national database that contains samples, not only from murderers and sex offenders, but even from burglars and car thieves.

Few people worry about the civil rights of convicted murderers or rapists, but critics do question how far these criminal databases should go. Four states (Virginia, Wyoming, New Mexico, and Alabama) have laws requiring submission of DNA samples from people convicted of any felony, not just violent crimes, and other states are considering such laws. A 1999 Louisiana law goes even further, requiring a sample from anyone who is arrested. In March 1999, Attorney General Janet Reno asked a federal commission to study the legality of such a requirement at the national level. A top government official made a similar proposal in Britain in 1998.

DNA samples have sometimes been required of people who were not even accused of a crime. For instance, the Department of Defense began collecting blood from all military personnel in 1992 for a DNA database that would be used, it claimed, to identify the remains of soldiers killed in action. There have even been proposals for a national database containing DNA records from all citizens. Groups such as the American Civil Liberties Union (ACLU) and the Libertarian Party see such a database as a major invasion of privacy. They question whether requiring DNA samples is constitutional, particularly if the samples are in the form of blood rather than hair or saliva. Taking blood requires an intrusion into the body, they say, and therefore might violate Fourth Amendment protection against unreasonable searches and seizures. If proposals for national DNA databases of noncriminal citizens go forward, they are sure to be hotly debated.

GENETIC HEALTH TESTING AND DISCRIMINATION

Leroy Hood, a pioneer in the development of automated genome analysis machines, said in 1987 that in the next century, "When a baby is born, we'll 'read out' his genetic code, and there'll be a book of things he'll have to watch for."[23] The Human Genome Project and new technology such as DNA chips (sometimes called gene chips or biochips), devices that use DNA on a computer chip to scan thousands of genes at a time for mutations that cause or increase the risk of disease, are bringing the future described in Hood's prediction rapidly closer. Some people wonder, however, whether such detailed knowledge of each individual's genes will be an entirely good thing. On the one hand, it could help people avoid illnesses to which their genes predispose them by, say, making certain lifestyle choices or undergoing tests for certain conditions unusually often. On the other hand, critics say, it could also lead to a new form of discrimination, based not on race or gender but on one's genes. "Genetic discrimination is the civil-rights issue of the 21st century," says Martha Volner, health policy director for the Alliance of Genetic Support Groups, an organization for families with inherited diseases.[24]

Potentially, as the ability to analyze individual genomes increases, genetic discrimination could affect anyone. "All of us have something or other in our genes that's going to get us in trouble," says Nancy Wexler of Columbia University, a leader in research on the form of inherited brain degeneration called Huntington's disease—and at risk for the disease herself because her mother had it. "We'll all be uninsurable."[25]

Genetic discrimination is most likely to limit access to life and health insurance, the latter of which, at least in the United States, is more or less required for quality health care. It could also bar people from jobs because most large employers provide health insurance for their workers, and employers may not want to risk having their group insurance premiums raised by hiring workers who seem likely to become sick. Genetic discrimination might even affect marriages and family relationships.

Discrimination on the basis of genetic makeup, as it was shown in characteristics thought to be inherited, existed before the word *gene* even entered the English language. This discrimination arose from the so-called science of eugenics (the word means "well born"), which was founded in the late 19th century by Francis Galton, a cousin of Charles Darwin. Galton believed that intelligence and other personality characteristics were inherited. He said that people with desirable characteristics (as he and his social group defined them) should be enouraged to have children, whereas people with undesirable characteristics (often meaning characteristics of other ethnic groups or social classes) should be discouraged or even forcibly prevented from doing so. In 1872 he wrote:

> *It may become to be avowed as a paramount duty, to anticipate the slow and stubborn process of natural selection, by endeavouring to breed out feeble constitutions, and petty and ignoble instincts, and to breed in those which are vigorous and noble and social.*[26]

Galton and his followers saw such practices as the human counterpart of scientific animal and plant breeding.

Around the start of the 20th century, Galton established the Eugenics Society in Britain to carry out his aims. Similar groups were formed in the United States and Germany and attracted many members. People in the American and European middle and upper classes saw eugenics as an answer to both the (perceived) prolific breeding of the lower classes, especially immigrants, and the economic costs of caring for the physically and mentally disabled. Eugenics supporters in the early years of the century included both respected scientists and such well-known nonscientific figures as Theodore Roosevelt, George Bernard Shaw, and (in his youth) Winston Churchill.

Negative eugenics—forcible blocking of reproduction for those thought not fit to reproduce—was codified into law in many places. By the 1930s, some 34 states in the United States had passed laws requiring the forcible sterilization of criminals, the mentally retarded (developmentally disabled), the insane, or others considered unfit to reproduce. According to one estimate, about 60,000 Americans were sterilized as a result of these laws. In the 1927 case of *Buck v. Bell*, the Supreme Court upheld the forced sterilization of a developmentally disabled woman, Carrie Buck. Noting that both Buck's mother and her seven-month-old daughter also seemed to have subnormal intelligence, renowned Justice Oliver Wendell Holmes wrote in his majority opinion on the case, "Three generations of imbeciles are enough."[27]

Denmark, Finland, Norway, Sweden, and some Canadian provinces also enacted eugenics laws in the 1920s or 1930s. Germany passed a eugenic sterilization law in 1933, soon after the Nazis took power, and used it to force sterilization of some 400,000 institutionalized people whose defects were presumed to be inherited. Later the Nazis carried eugenics to its extreme by trying to wipe out the genes of whole ethnic groups through mass killing.

Partly because of the Nazi excesses, the tide of public opinion began to turn against eugenics around the time of World War II. In *Skinner v. Oklahoma* in 1942, for instance, the Supreme Court struck down a state law requiring forced sterilization of convicted criminals, saying that procreation is "one of the basic civil rights of man."[28] Belief in eugenics never totally died out, however. Many eugenics laws remained on the books until the 1970s, although they were seldom enforced. As late as 1980, 44 percent of respondents in a United States poll favored compulsory sterilization of habitual criminals and the incurable mentally ill. Even today, China has a law that can deny permission for marriage to or force sterilization or abortions on people with certain inherited diseases.

In the 1970s, just as traditional eugenics was fading, a new form of genetic discrimination began to appear. Tests that examine DNA directly were still decades away, but sufferers from and carriers of certain inherited diseases could be identified indirectly by tests for particular substances in their blood or other body fluids. Chief among these illnesses was sickle-cell disease (also called sickle-cell anemia), a blood disease that affects mostly people of African ancestry.

Sickle-cell disease is caused by a single recessive gene that codes for an abnormal form of hemoglobin, the protein in red blood cells that gives them their color and allows them to carry oxygen through the body. Cells containing the abnormal hemoglobin take on a crescent shape and block the body's smallest blood vessels, depriving tissues of oxygen and causing considerable pain, disability, and sometimes death. Only people who inherit defective genes from both parents suffer from the disease, however. Those who inherit a defective gene from one parent and a normal one from the other—called sickle-cell carriers or possessors of the sickle-cell trait—can pass the disease to their children, but they themselves are quite healthy. Indeed, they may have had a health advantage in Africa, because sickle cell trait seems to be associated with resistance to malaria, an endemic disease in much of the continent. About one in 500 African Americans suffers from sickle-cell disease, and one in ten is a carrier.

Tests can detect sickle-cell disease and sickle-cell trait by examining hemoglobin in blood samples, and in the early 1970s African Americans were widely screened for this condition. The screening programs were intended to identify carriers so they could be counseled about the risk of having children with the disease if they married other carriers. Because of widespread lack of understanding about the difference between sickle-cell carriers and people with the full-blown disease, however, some carriers were charged high insurance premiums or denied certain kinds of employment. The Air

Force and some private airlines, for instance, barred sickle-cell carriers from being pilots because of an unproven belief that such people would have trouble with the decreased oxygen levels at high altitudes.

African Americans supported the sickle-cell screening programs at first, but they soon began to protest them as racism in a new, medical guise. A number of states therefore either dropped the programs or passed laws limiting use of information obtained from them. North Carolina, for instance, passed a law in 1975 that prohibited employers from discriminating against anyone possessing sickle-cell trait or the type of hemoglobin associated with it. This was the first American law to address genetic discrimination in the workplace. Four other states passed similar laws in the next decade. The federal government, for its part, passed the Sickle-Cell Anemia Act in 1972 and the Genetic Diseases Act in 1976. These mandated, among other things, that screening be voluntary and unlinked to eligibility for federal services.

The same issues raised by sickle-cell screening in the 1970s reappeared in the 1990s as a growing number of tests became available to screen DNA for mutations that cause or increase the risk of particular diseases. By early 1999, genetic centers could test for 30 to 40 inheritable conditions, in some cases with an accuracy of 99 percent. These included not only classic inherited diseases such as Huntington's but some forms of more common and complex illnesses, such as cancer and heart disease.

Critics have pointed out a number of problems with the current generation of genetic tests. First, the tests' results are often hard to interpret and even harder to explain to tested people and their families. Tests and testing laboratories are also poorly regulated, and results may not be accurate. In many cases, too, the tests do little to improve health because people and their physicians cannot prevent or cure the diseases that the tests predict. Finally, the tests make people focus on individual genes as a cause of disease when, in fact, most illnesses—including the most common and deadly ones, such as cancer and heart disease—result from the interaction of multiple genes with each other and with environmental factors such as lifestyle choices (smoking, eating a high-fat diet) or exposure to toxic chemicals. Thus, even if a person inherits a gene that increases the risk of, say, heart disease, that person may never develop the disease, whereas someone else with a better genetic profile but a worse assortment of environmental exposures may.

Chief among the legal issues raised by genetic tests are invasion of privacy (an important issue with regard to medical records in general) and discrimination, especially in insurance and employment. One recent invasion of privacy case involved Lawrence Berkeley Laboratory (LBL), a California research facility funded by the Department of Energy. In February 1998, the U.S. Court of Appeals for the 9th Circuit ruled that LBL had been wrong to test the blood of its employees for sickle-cell trait and other conditions without their informed consent. The court held that such action violated the protection against unreasonable search and seizure guaranteed by the Fourth Amendment. Because some tests were performed only on samples from blacks, Hispanics, or women, they also violated the 1964 Civil Rights Act. This was said to be the first case in which a federal appeals court recognized a constitutional right to genetic privacy.

National government DNA databases such as the ones established in 1992 by the Department of Defense and in 1998 by the FBI have also been seen as threats to genetic privacy. In April 1996, John C. Mayfield III and Joseph Vlakovsky, both Marines with exemplary military records, were convicted in a court martial of disobeying a direct order because they refused to give blood for storage in the Department of Defense database. The soldiers said that letting the government archive their genetic information violated their right to privacy, and they feared that the information might eventually be used to discriminate against them in some way. This was the first case to challenge the right of an employer to mandate sample donation for genetic testing.

Many people have told interviewers that they are afraid to be tested for defective genes they suspect they carry because they fear that insurers or employers will gain access to the results and then deny them or members of their families insurance and jobs. A survey made in the late 1990s, for example, showed that nearly 70 percent of the respondents were somewhat or very concerned that genetic information might be used against them by either their health insurers or their employers. Of 279 women with high rates of breast and ovarian cancer in their families who were surveyed in 1996, only 43 percent wanted to be tested for mutations in BRCA1, a gene associated with familial risk of these cancers. Many of those who chose not to take the test said they did so because of fear of discrimination.

Several studies suggest that such fears are justified. One published in 1996 by Paul Billings, then at Stanford University, cited 455 cases of people being denied health care, insurance, jobs, schooling, or the right to adopt children because of a family history of inherited disease. A Harvard study done in the same year found more than 200 examples of healthy people who believed they had been discriminated against because of inherited disease in their families or their own genetic background. A third 1996 study, in which members of support groups for families with 101 different inherited diseases were polled, reported that 25 percent of the 322 respondents believed they or family members had been refused life insurance, 22 percent believed they had been refused health insurance, and 18 percent felt they had lost jobs because of their perceived health risks.

People's perceptions of discrimination, of course, may or may not be accurate. Dean Rosen, senior vice president of policy for the Health Insurance Association of America, insisted in early 1999 that "the fears out there are just not reality."[29] But another insurance executive, who preferred to remain anonymous, advised people to "apply for insurance today, get [genetically] tested tomorrow" rather than the other way around.[30]

More than 30 states have enacted laws prohibiting insurers or employers from requiring applicants to take genetic tests, denying these groups access to genetic information, or preventing them from using such information in their decisions. Most other states are considering such laws. Laws of this kind, however, are often weakened by lack of a clear definition of genetic testing and, therefore, the kind of information that is covered. The laws also often do not cover genetic information that insurance companies obtain indirectly, for instance through family histories or through tests for gene products such as abnormal hemoglobin in the blood of people with sickle-cell trait. Finally, state insurance laws do not affect employers' self-funded plans, which in 1995 provided insurance for over a third of the nonelderly insured population.

In his State of the Union speech in January 1998, President Clinton called for laws to prevent discrimination on the basis of genetic information. Vice President Al Gore made a similar request at about the same time, referring specifically to discrimination in the workplace. A number of federal laws banning genetic discrimination have been proposed—some 20 such bills were before Congress in early 1999—but as of that time, none of them had passed. In fact, a bill that passed the House of Representatives in July 1999, although it protected consumers' privacy by forbidding banks to reveal data about them, also contained a provision that allowed health insurers to reveal genetic and other medical information to credit card companies.

Meanwhile, two existing federal laws do, or at least may, have an impact on genetic health testing and discrimination. The Health Insurance Portability and Accountability Act, passed in 1996, is aimed mainly at preventing people from losing their health insurance when they change jobs, but it includes a provision that forbids health insurers issuing group plans to deny insurance on the basis of preexisting genetic conditions. It does not cover the 5 to 10 percent of Americans who have individual insurance plans, however, and it may not cover companies that insure their own workers, as opposed to those who buy group plans from insurance companies. It also does not necessarily prevent insurers from raising premiums on the basis of genetic information or protect the privacy of such information.

The Americans with Disabilities Act, passed in 1990, may also apply to healthy people who have inherited "bad" genes, if they can show that the results of genetic tests have made their employers perceive them as disabled. The federal Equal Employment Opportunity Commission ruled in 1995 that denying employment to people who are healthy but have a genetic predisposition to a disease is illegal under the ADA. The EEOC ruling so far has not been tested directly in the courts.

Biotechnology and Genetic Engineering

In *Bragdon v. Abbott,* a case brought before the Supreme Court in 1998, the court ruled that the plaintiff, who had tested positive for HIV but did not yet show any symptoms of AIDS, was protected by the ADA. A person carrying, say, BRCA1 or the gene for Huntington's disease would seem to be in a similar situation. In a June 1999 Supreme Court ruling on several other ADA cases, however, Justice Sandra Day O'Connor wrote in her majority opinion, "We think the language [of the ADA] is properly read as requiring that a person be presently—not potentially or hypothetically—substantially limited [in a major life activity] in order to demonstrate a disability."[31] This would seem to dim hopes that the ADA will cover healthy people with genetic predispositions to illness.

Discrimination based on genetic health testing has also been an important issue in Britain, where the national health care system is increasingly supplemented by private insurance. The government's Human Genetics Advisory Commission recommended in 1997 that there be a two-year moratorium on disclosure of the results of genetic testing and that a mechanism be established to evaluate the scientific evidence in support of particular genetic tests. In January 1998 the Association of British Insurers issued a code of practice on handling genetic information that named seven diseases in which the results of genetic tests had to be revealed to insurers. The code also said that, although insurers would ask for family histories and results of genetic tests, they would not take them into account for insurance policies under 100,000 pounds sterling that were associated with mortgages on the principal residence of the insured. They would consider such information for other insurance or larger mortgage policies, however.

Efforts have also been made to protect against genetic discrimination in other European countries and in Europe as a whole. Norway has laws regulating genetic testing, for instance. Article 11 of the Convention for Human Rights and Biomedicine, adopted by the Council of Europe in November 1997, prohibits discrimination on grounds of genetic heritage, and the UNESCO General Conference's Universal Declaration on the Human Genome and Human Rights, also adopted in 1997, has a similar provision. As with laws opposing genetic discrimination in the United States, however, the meanings of terms such as *genetic testing* are not clearly defined in these rulings. The methods by which the declarations are to be enforced are also not clear.

The debate about insurers' use of genetic information is part of a basic conflict between insurers (and employers who pay for employees' insurance) and insured regarding access to and use of all medical information. As a number of commentators have pointed out, this conflict is built into the nature of life and health insurance. Insurers can stay in business and make a profit only if they and the people they insure have equal knowledge of those individuals' risks of illness or death. If insured people alone know about a risk, the insurers say, they may take out large amounts of health or life insurance because they expect to need it. This "adverse selection" throws off the statistical methods by which insurance premiums are determined and drives up premiums for everyone. Harvie Raymond, director of managed care and insurance operations at the Health Insurance Association of America, insisted in 1995 that "he who assumes a risk should have the opportunity to evaluate that risk" and that laws barring insurance companies from obtaining genetic information "could create a great deal of havoc in the industry, causing costs to go up and fewer people to ultimately be able to afford coverage."[32] Conversely, if insurers know of a person's increased risk, they have a powerful financial incentive to deny insurance or charge very high premiums to that individual. This could make insurance unavailable to the very people who need it most.

Some commentators believe that, as genetic risks become easier to predict, this conflict will become soluble only by separating health care from the insurance system, probably by turning to a national health care system like the ones in Britain and Canada. Lori B. Andrews of the Chicago–Kent College of Law has predicted that "increasingly sophisticated genetic diagnostic tests may force a total rethinking of the concept of health insurance" and that the perceived injustice of genetic discrimination "will provide the impetus for the development of a national health system."[33] Another possibility, suggested by genetics expert Thomas Caskey, is to place people at high risk of genetic disease in a special insurance pool, as is sometimes done with otherwise uninsurable drivers.

Another form of discrimination could stem from genetic determinism, the widely held belief that genetics—perhaps even single genes—are primarily or wholly responsible for such behaviors as violence, risk taking, and homosexuality. Although these behaviors may have a genetic association in certain cases, they are most likely determined by complex interactions between genes and environment. Nonetheless, if genetic determinism continues to be popular and genes associated with, say, violence are found, civil libertarians say, children with such genes might be put in special schools or otherwise segregated or treated differently from other children. Attributing violent behavior to genetics could also lead to a new variant of racism if "violence genes" prove to be more common in some racial or ethnic groups than in others. Perhaps the most likely and also most important danger of focusing on exclusively genetic causes of behavior such as violence, critics say, is the diversion of attention and funding from correction of social causes of such behavior. "We know what causes violence in our society," says genetic discrimination expert Paul Billings. "[It is] poverty, discrimination, [and] the failure of our educational system."[34]

HUMAN GENE ALTERATION AND CLONING

On September 14, 1990, a four-year-old girl watching *Sesame Street* from her hospital bed made history. The child, Ashanthi deSilva, had been born with a mutant gene that made her body unable to produce an enzyme called adenosine deaminase (ADA). Lacking this protein, some of the cells in her immune system could not thrive. As a result, like other people with poorly functioning immune systems, she was easy prey for every microbe to which she was exposed. During all of her short life she had suffered from one infection after another.

Life changed for Ashanthi—and, potentially, for the world—on that September day. W. French Anderson, Michael Blaese, Kenneth Culver, and their coworkers at the National Institutes of Health (NIH) had devised a technique for inserting a normal ADA gene into white cells (part of the immune system) taken from blood. When blood cells extracted from her earlier and treated with this method were reinjected into Ashanthi as she watched her hospital television, she became the first person to receive gene therapy, or injection of altered genes for the purpose of treating disease.

Ashanthi's treatment was not the first attempt at gene therapy. A UCLA physician named Martin Cline had asked the Recombinant DNA Advisory Commission (RAC) for permission to try such therapy for thalassemia, another inherited blood disease caused by a single mutant gene, in 1980. When he was refused, he left the country and tried the treatment on patients in Italy and Israel. Cline failed even to get genes into his patients, let alone help them, and his failure gave gene therapy a bad name. Thus it was no surprise that when Anderson first asked permission to try gene therapy for ADA deficiency, as Ashanthi's disorder was known, in 1987, the RAC turned him down as well, even though he produced more than 500 pages of documentation showing the treatment's safety and likely success.

Deciding to take another approach, Anderson joined forces with another NIH scientist, Steven Rosenberg, in 1988. Rosenberg, then chief of surgery at the National Cancer Institute, was trying to devise ways to boost the power of certain immune system cells to attack cancer. As would later be done with Ashanthi, he withdrew blood from his patients, treated it in the laboratory, and then reinjected it. His treatment itself did not involve genetic engineering, but Rosenberg wanted to insert a harmless gene into the altered cells so that he could more easily identify and track them after injection.

Rosenberg's proposed use of altered genes had a better chance of winning approval than Anderson's, both because the genes themselves were not expected to have any medical effect and because the patients involved were all expected to die from their cancers within a few months. Nonetheless, the

RAC, spurred in part by protests from the ubiquitous Jeremy Rifkin, also rejected Rosenberg's proposal several times. It finally granted approval, however, and on May 22, 1989, a 52-year-old truck driver with advanced melanoma (a fast-growing, deadly form of skin cancer) became the first human to have genetically altered cells placed in his body. As expected, the inserted gene, which coded for resistance to a certain antibiotic, did not affect his cancer. It also did no identifiable harm to the man's overall health.

The success, at least from a safety point of view, of Rosenberg's treatment paved the way for eventual approval of Anderson's treatment of Ashanthi. Anderson nonetheless had to undergo months of questioning by seven different committees before he succeeded in winning permission from both the RAC and the FDA. Some of the hardest questions concerned the safety of the treatment, which involved the use of former cancer-causing viruses that had been altered to make them unable to cause disease. Viruses, nature's own genetic engineers, reproduce by inserting their genes into the genomes of the cells they infect. The cells then reproduce the viral genes along with their own. Anderson's group had placed a normal human ADA gene in these viruses, and they hoped that the viruses in turn would insert that gene into Ashanthi's white cells.

Ashanthi's treatment had to be repeated several times during the next two years, but with it she thrived, changing from a sickly, house-bound toddler to, as French Anderson reported five years later, "a, healthy, vibrant nine-year-old who loves life and does *everything.*"[35] A similar treatment given to another girl with ADA deficiency was equally successful. Although the girls' disease cannot be cured, the treatments have controlled it well enough to let them lead nearly normal lives. In early 1999 they were still healthy.

Since then, gene therapy has been tried experimentally on a host of inherited disorders, and scientists are working on gene treatments for more common diseases that have a genetic component, such as cancer and AIDS. By late 1996, about 1,500 people worldwide had received changed genes as part of more than 200 trials aimed at about 30 different diseases. To be sure, not all the treatments have succeeded as well as Ashanthi's, and none has completely cured a disease; getting the genes into enough cells to do the patient good and keeping both the cells and the genes active remain difficult problems. On the other hand, the experimental treatments have seldom worsened any patient's condition, although in 1999 an Arizona man was said to have died as a result of a gene therapy experiment. All new gene therapy proposals must be approved by the FDA, but only those that are substantially different from previous ones—those that use a different type of virus to insert the genes, for example—need to be reviewed by the RAC.

The potential social effects of human gene therapy may prove at least as important as its medical effects. Both supporters such as French Anderson and critics such as Jeremy Rifkin agree that Ashanthi deSilva's treatment represented not only a scientific advance but, as Anderson said at the time, "a cultural breakthrough, . . . an event that changes the way we as a society think about ourselves."[36] They and many others, however, are still arguing about whether that change is for good or ill.

Few people question the morality of altering genes of human body cells to prevent or cure a life-threatening illness like Ashanthi's (although Germany, perhaps sensitive about the eugenics aspect of its Nazi past, forbids any alteration of human genes). After that, however, the ethical ground becomes shakier. Rifkin and other critics fear that, once the human genome has been completely decoded and the technical problems that presently limit human gene alteration have been solved, the definition of "disease" or "defect" will be stretched to include relatively minor problems (such as nearsightedness or obesity) or even mere differences (such as shorter-than-average height). If all these are engineered away, the result could be, at best, an undesirable loss of genetic diversity or, at worst, a new form of eugenics. Even French Anderson has said he feels strongly that gene therapy should be used only to treat serious disease.

Although governments might insist on, or at least strongly encourage, alteration of "defective" genes as a way to control health care costs, many of both critics and supporters of human gene alteration suspect that the demand for gene changes is more likely to arise from market forces than from government fiat. In addition to removing what they see as deficiencies, well-to-do parents might try to create "designer babies" by inserting genes likely to produce, say, intelligence or physical beauty, just as they now purchase orthodontic treatments or special schooling for their children.

The ethical questions raised by human gene alteration become especially great if germ-line genes—those in the sex cells (the cells that become sperm and eggs), whose genetic information is passed on to offspring—are altered. Germ-line gene alteration has not yet been attempted in humans, although it has been done in animals; all present human gene therapy treatments affect only somatic (body) cells and produce no changes that can be inherited. Critics say that, for one thing, altering germ-line cells would violate the rights of the unborn because they cannot give consent to it. Furthermore, even removal of a gene known to be associated with a serious disease might not be as clear-cut a good as it might seem, as the relationship between sickle-cell trait and resistance to malaria shows. In permanently deleting such a gene, therapists might unknowingly delete some characteristic that the human species will need at a future time.

At present, most people, including most scientists in the field, seem to feel that the human germ line should never be changed. Fifty-six religious leaders presented a statement to that effect to Congress in 1986, for instance, and

Jean Dausset, founder of the Human Polymorphism Study Centre on Paris, made a similar statement in the September 1994 *UNESCO Courier*. No law actually forbids such action, however, and treatments that approach it may soon take place. In late 1998, French Anderson warned the RAC that in two or three years he would ask for approval to use gene therapy on fetuses that carry certain inherited defects, including ADA deficiency. Although unlikely and not part of Anderson's plan, changed genes inserted by this sort of therapy could get into the fetus's sex cells and produce a change in its germ line.

The form of human gene alteration that has generated the most heated debate in the late 1990s is cloning. The possibility of cloning humans has haunted movies and novels as well as ethical discussions ever since genetic engineering began. For example, *In His Image: The Cloning of a Man*, a 1978 book by science writer David Rorvik, caused an uproar because it claimed that Rorvik had witnessed the cloning of an unnamed wealthy man, although the book proved to be almost surely fictional. Movies featuring human cloning have included Woody Allen's *Sleeper* (1973), *The Boys from Brazil* (1978), *Multiplicity* (1996), and *Gattaca* (1997).

Human cloning moved closer to reality in October 1993, when Robert Stillman and Jerry Hall of George Washington University Medical Center in Washington, D.C., announced that they had separated the cells of very early human embryos and made them multiply, producing 48 embryos from a starting batch of 17. They used defective embryos already scheduled to be discarded by a fertility clinic, and they did not allow the embryos to develop into fetuses, but their work nonetheless caused a considerable outcry. An even greater furor followed announcement of the cloning of Dolly the sheep from an adult ewe cell in February 1997 because it raised the possibility that human adults could produce duplicates of themselves.

Objections to human cloning range from the practical one that, at least at present, such a procedure would be very risky to the prospective fetus (Dolly's was the only live birth out of 277 tries) to the quasi-religious belief that cloning humans would be "playing God." Critics also fear that the parents of cloned children, or the children themselves, would regard the children as mere "carbon copies" or "products" rather than independent human beings. Originators of clones might either expect to control the cloned children's lives completely or, conversely, disclaim all responsibility for them, saying that they were siblings rather than true children. Some opponents of human cloning have pictured nightmare scenarios featuring, at one extreme, cadres of identical Hitlers (or, at best, Madonnas or Michael Jordans) or, at the other, armies of mindless slaves or even warehouses of headless bodies kept for possible organ donation.

Supporters of human cloning say that cloning could provide help to infertile couples who can reproduce in no other way. They point out that human cloning already occurs naturally in the form of identical twins, and they claim that an artificially produced human clone would simply be an age-delayed twin. They say that such a clone would be just as much a separate individual, with his or her own personality, as a twin is; indeed, the personality differences between a clone and its original would be greater than those between twins because they would be raised in different eras and therefore would be bound to have very different life experiences. Clones would be entitled to all the legal rights that other citizens have, including protection from slavery or forced organ donation.

As has happened with previous outcries against new forms of biotechnology, public fear was followed by increases in regulation and attempted changes in law. On March 4, 1997, less than a month after Ian Wilmut's article describing the cloning of Dolly appeared in *Nature*, President Clinton called for a ban on the use of federal funds for research on human cloning. "Any discovery that touches upon human creation is not simply a matter of scientific inquiry," he said. "It is a matter of morality and spirituality as well."[37] He also ordered the National Bioethics Advisory Commission (NBAC), a group of 18 experts in theology, science, and law headed by Princeton University president Harold Shapiro, to study the issue for 90 days and provide recommendations.

A day later, Congress held its first hearings on cloning, which took place in a meeting of the House Science Committee's Technology Subcommittee. During those hearings, even the Biotechnology Industry Organization, the industry's chief lobbying group, and Dolly creator Wilmut agreed that full development of cloned humans should be banned, at least for the time being. NIH director Harold Varmus, however, warned against legislating too broadly or hastily: "Legislation and science don't mix very well," he said. "Legislation is difficult to reverse."[38] Similarly, Wilmut warned against banning all cloning research, some of which he said held great promise for medicine; he urged Congress not to "throw out this particular baby with the bath water."[39]

In spite of such cautionary notes, Senator Christopher Bond (R-Mo.) quickly introduced a Senate bill to ban federal funding for human cloning research, including research on cloned embryos that would not be allowed to develop. Congressman Vernon Ehlers (R-Mich.) introduced two similar bills into the House of Representatives, one of which would have banned federal funding for human cloning research and the other of which would have banned all such research regardless of its funding source. The funding bill passed the House Science Committee in July 1997.

The NBAC presented its report to President Clinton in June 1997. The commission recommended that all attempts to create fully developed humans through somatic cell nuclear transfer cloning (the method used to create Dolly) be banned for three to five years, primarily because the experimental procedure would present unacceptable risks to the fetus. On the other hand, it concluded that research with cloned human DNA sequences, cell lines, and animals should be allowed to continue.

Public outcry against human cloning died down somewhat in the second half of 1997, but it revived in January 1998 when Richard Seed, a Chicago physicist, announced that he planned to set up a clinic to clone children for infertile couples, using the Dolly method, within the next year. Most scientists doubted that Seed actually had the ability to produce human clones, but his claim nonetheless made headlines. The FDA said it would not grant permission for any such activity in the United States. Within the month, California passed a law forbidding human cloning, becoming the first state to do so, and other states seemed poised to follow.

The federal government also renewed its interest in cloning bans. In his yearly State of the Union Address on January 27, President Clinton called for legislation to ban human cloning, and several members of Congress obliged. Senator Bond's bill was reintroduced on February 3 as S.1601, with the support of powerful Senate Majority Leader Trent Lott (R-Miss.) and Bill Frist, a Republican from Tennessee who was the Senate's only physician member. The bill was put on a "fast track" and sent to the Senate floor without having to pass through a committee first. Lobbying by biotechnology and medical research organizations, patient advocacy groups, and individual scientists (including 27 Nobel Prize winners) blocked it at that point, however. The biotechnology industry preferred a rival bill, S. 1602, which was cosponsored by Ted Kennedy (D-Mass.) and Dianne Feinstein (D-Calif.). That bill would ban human cloning for at least 10 years but would permit research with undeveloped embryos, which scientists said had great potential value in producing sources for transplantable tissues.

As of late 1999, no federal bill banning human cloning has become law. Even if Congress does pass such a bill, some legal scholars have doubted whether the resulting law would be constitutional. The Supreme Court has defended procreation as a basic civil right, and the U.S. District Court for the Northern District of Illinois stated in the 1990 case of *Lifchez v. Hartigan* that this right includes "the right to submit to a medical procedure that may bring about . . . pregnancy," such as in vitro (test tube) fertilization—or, probably, cloning.[40] A right to scientific inquiry may also be included within the rights of free speech and personal liberty, although Congress has been able to restrict the methods by which such inquiry is conducted, for instance by requiring informed consent from human experimental subjects. Even if a ban on human cloning withstood court challenges, it would be difficult to enforce because the technology likely to be used is relatively simple and would not require elaborate facilities or large amounts of funding.

The United States is far from the only country to consider or even pass laws against human cloning. The World Health Organization (WHO) and the World Medical Association both issued statements opposing the cloning of human beings in mid 1997. The Council of Europe, representing 19 European nations, signed an agreement banning human cloning on January 12, 1998. A number of individual nations, including Australia, Belgium, Britain, Denmark, Germany, Israel, the Netherlands, and Spain, have also banned or at least placed a moratorium on cloning human beings. At least for now, the supporters of cloning, at least of fully developed humans, are definitely in the minority.

FUTURE TRENDS

No one knows what advances the coming "biotech century" (as Jeremy Rifkin has dubbed it) will bring, but they are sure to be amazing, and their effects on society will be challenging, if not wrenching. There were good reasons why James Watson, the first head of the Human Genome Project, earmarked 3 percent of the project's budget for investigation of its ethical, legal, and social implications (ELSI). The implications of developments in biotechnology affecting plants and animals will be equally powerful.

In agricultural biotechnology, Monsanto Corporation CEO Robert Shapiro claims that genetically engineered crops and other advances in agricultural biotechnology will provide a "sustainable" way to nourish the world's rapidly growing population, simultaneously increasing both the quantity and the nutritional value of food and preserving the environment by decreasing soil erosion, energy use, and dependence on pesticides and other chemicals. Transgenic plants and animals may also become inexpensive sources of vaccines and medicines, thus improving world health, and of industrial products such as plastics. Critics such as Jeremy Rifkin and Rural Advancement Foundation International (RAFI), however, fear the effects of having this new technology increasingly concentrated under the control of international giants like Monsanto. They say that most farmers in developing countries are likely to be either shut out of the genetic bonanza or forced to be totally dependent on large agrochemical companies.

Some environmentalists are also concerned about unexpected damage that genetically engineered crops may do when they become widespread, such as becoming (or causing related wild plants to become) unkillable weeds or harming beneficial insects or other wildlife. At very least, they say, such crops are likely to accelerate the loss of biodiversity that has already been produced by monoculture, the practice of growing huge numbers of nearly identical plants. Especially in Europe, many people fear that genetically modified food crops will prove to harm human health as well as the environment. Threats, or at least fears, may increase as more altered crops come to contain genes from distant plant species or even animals. Because fears of genetically altered crops are greater in some parts of the world than others, trade conflicts such as the current ones over whether genetically modified foods should be labeled are likely to continue.

Legal and trade conflicts are likely to continue over patents and other protection of intellectual property in biotechnology as well, particularly in regard to biologically useful material discovered in developing countries and prepared for market by large companies in industrialized nations. Biotechnology companies, for their part, feel that they need and deserve patent protection in exchange for the considerable time, effort, and money they must spend in bringing new products to market, and they fear having their work exploited without compensation in countries where intellectual property laws are weak. They are sure to continue to demand the strengthening of such laws in international treaties and trade agreements.

The biotechnology companies' feelings of mistrust are certainly shared by the tropical countries from which the raw materials for many new medicines and other products come. Representatives of those countries say that large multinational companies exploit their natural resources and the development work that their native farmers and healers have carried out for centuries. They accuse research efforts such as the Human Genome Diversity Project of exploiting the native people themselves, taking their cells or genes and commercializing them (or allowing others to do so) with little or no compensation. Organizations such as RAFI and Cultural Survival also claim that researchers from industrialized countries frequently disregard other cultures' views about such subjects as the human body and ownership of living things.

Within industrialized countries, too, differing opinions about the morality of modifying and of patenting or "owning" types of living things, body parts, cells, and genes will surely continue to clash. Attitudes of religious groups to these things differ, but many such groups feel that some activities of biotechnologists usurp the role of God. Numerous polls indicate that, even among people who do not have strong religious beliefs, large numbers feel a deep moral unease about these subjects. The moral unease is deepest when the materials being patented or changed are human. This unease can only increase as the flood of information released by the Human Genome Project and related research grows.

As the complete sequences of more and more genomes are determined, the focus of biotechnology is shifting to the management and sale of genetic information. This focus has spawned new scientific disciplines called genomics and bioinformatics, which marry genetics and computer technology. These new sciences will increase the efficiency and scope of genetic engineering, for instance speeding the development of drugs and allowing the entire metabolisms of plants, animals, and even perhaps humans to be radically altered. They will also raise new ethical issues as some scientists or businesspeople begin to see (or to be accused of seeing) living things as mere bundles of information, as open to modification as a computer program.

Disputes about the validity of DNA identification in court cases will probably continue to die down as the technology improves, particularly if standards for police evidence gathering and laboratory testing procedures can be agreed upon and enforced. Increased automation of tests and interpretation should help in removing the threat of human error.

Arguments about genetic testing in relation to health care, on the other hand, are sure to increase as such testing becomes more detailed, accurate, and widespread. Unless some way to slow the rise in health care costs can be found, conflicts between those who demand the latest health care technology and those who are expected to pay for it (whether they are insurers, employers, government, or some combination of these) will grow more acute. Federal laws forbidding genetic discrimination probably will be passed, but enforcing them may be difficult.

Solving the technical problems besetting gene therapy may take several decades, but eventually alteration of genes in human body cells, probably increasingly done before birth, is likely to make inherited disease a thing of the past. It is also likely to be a part of treatments for more common conditions such as cancer and heart disease. "Health" and "disease" are parts of a continuum, however, and no one knows what degree of gene alteration society will prove willing to accept. At least some wealthy parents, clandestinely if necessary, will probably attempt to enhance the genetic endowment of normal children, but they may discover that producing another Einstein or Marilyn Monroe is easier said than done because environment shapes a child as much as genetics. If gene enhancement becomes accepted and common, people with unaltered genes may come to be looked down upon or, conversely, to demand gene alteration as their (or their children's) birthright.

There is no technical reason why germ-line genes should be any more difficult to alter than the genes of body cells, but arguments about the ethics of making such alteration in humans are guaranteed to be more acute than those surrounding gene therapy or other alteration of body cell genes. Germ-line alteration, after all, goes beyond the individual to potentially affect the evolution of the entire species. Many people are sure to doubt the wisdom of taking on such an awesome responsibility.

Human cloning will also continue to be a sore subject. Whether or not such cloning is legally banned, someone, somewhere, is almost sure to produce a cloned human child within a decade or so. Fears may die down when people realize that a cloned child is not much different from any other baby, as happened with in vitro fertilization once it became widespread. Some of the nightmare scenarios will fade when people realize that clones are not instant adults (as in some movies), automatons, or exact personality duplicates of their "parents."

On the other hand, even if society comes to accept human cloning—which is by no means certain—such activity seems unlikely to become common. As a way to produce armies of either dictators or slaves, or even as a way to "resurrect" a lost loved one, most commentators agree, cloning simply will not work. It will probably be used only by a small number of infertile (including same-sex) couples and, perhaps, a few eccentrics who do not understand the interplay between genetics and environment. Laws defining family relationships will probably have to be modified if cloned human children come into existence, however, just as they have been altered in the past to accommodate surrogate motherhood, anonymous sperm donation, and other new reproductive technologies.

A new line of medical research is likely to do away with one of the most frightening human cloning scenarios, the production of clones (brainless or otherwise) as sources for tissue and organ transplants. At the same time, this research will be a strong force for allowing the cloning of human embryos that do not develop. The research grows out of the isolation in 1998 of human embryonic stem cells, which have the potential to grow into any type of cell in the body and therefore could supply immune-compatible tissues for any need. The most likely sources for stem cells at present are embryos discarded by fertility clinics or aborted fetuses. Alternatively, but more controversially, embryos could be cloned for stem cell harvesting, possibly from mature body cells of a person needing donated tissue. Scientists have experimentally cloned such cells in cow eggs (used because they are easily obtainable) from which the nucleus has been removed.

Federal funding cannot now be used for research that results in the destruction of human embryos, which includes stem cell research. Private companies can conduct such research, however, and several are doing so. To keep government-funded scientists from being left out of this very promising new field, both the NIH and the National Bioethics Advisory Commission have recommended allowing government funding to be used for research on stem cells.

The one thing that seems clear regarding the future of biotechnology and its regulation is that the public—not to mention the public's media sources and political representatives—will need great improvements in education to deal with the ethical challenges of the biotech century. People must learn to grasp such things as the statistics of probability and risk or the complex interaction between genes and environment. Only education can help people sort through the hype and the nightmares presented by supporters and opponents of various techniques and reach reasoned conclusions about how humanity should use its wonderful and terrible new power to alter the essence of life.

1 Unknown scientist, quoted in Edward Shorter, *The Health Century* (New York: Doubleday, 1987), p. 238.

2 Michael Rogers, quoted in James D. Watson and John Tooze, *The DNA Story* (San Francisco: W. H. Freeman, 1981), p. 28.

3 Paul Berg et al. (Asilomar Conference), quoted in N. A. Tiley, *Discovering DNA* (New York: Van Nostrand Reinhold, 1993), p. 258.

4 Norton Zinder, quoted in Burke Zimmerman, *Biofuture* (New York: Plenum Press, 1984), p. 141.

5 1978 NIH guidelines revision, quoted in Watson and Tooze, *The DNA Story*, p. 431.

6 Jeremy Rifkin, quoted in Paul Ciotti, "Saving Mankind from the Great Potato Menace," *California Magazine*, October 1984, p. 97.

7 Unknown newspaper story, quoted in Joel Gurin and Nancy E. Pfund, "Bonanza in the Bio Lab," *Nation*, November 22, 1980, p. 548.

8 USC 35, Section 103(a).

9 P. J. Federico, quoted in *Diamond v. Chakrabarty*, 447 U.S. 303.

10 Statement of Foundation on Economic Trends and General Board of Church and Society of the United Methodist Church, quoted in Ronald Cole-Turner, "Religion and Gene Patenting," *Science*, October 6, 1995, p. 52.

11 Jerry Caulder, quoted in Richard Stone, "Sweeping Patents Put Biotech Companies on Warpath," *Science*, May 5, 1995, p. 656.

12 European Parliament, quoted in Nigel Williams, "European Parliament Backs New Biopatent Guidelines," *Science*, July 25, 1997, p. 472.

13 *John Moore v. Regents of California*, 51 Cal. 3d 120.

14 Unknown juror, quoted in Peter J. Neufeld and Neville Coleman, "When Science Takes the Witness Stand," *Scientific American*, May 1990, p. 46.

15 Jamie Gorelick, quoted in "Fugitive Justice," *Nation*, March 3, 1997, p. 4.

16 R. C. Lewontin and Daniel L. Hartl, "Population Genetics in Forensic DNA Typing," *Science*, December 20, 1991, p. 1745ff.

17 Daniel Hartl, quoted in Leslie Roberts, "Fight Erupts over DNA Fingerprinting," *Science*, December 20, 1991, p. 1721.

18 *Frye v. United States*, quoted in Ricki Lewis, "Genetics Meets Forensics," *Bioscience*, January 1989, p. 6.

19 Eric Lander, quoted in Peter Aldhouse, "Geneticists Attack NRC Report as Scientifically Flawed," *Science*, February 5, 1993, p. 755.

20 David Kaye, quoted in Karyn Hede George, "DNA Fingerprinting Gets a Reprieve," *Technology Review*, November–December 1996, p. 16.

21 Daniel Hartl, quoted in Rachel Nowak, "Forensic DNA Goes to Court with O. J.," *Science*, September 2, 1994, p. 1353.

22 Ronald Allen, quoted in Jerry Adler and John McCormick, "The DNA Detectives," *U. S. News & World Report*, November 16, 1998, p. 66.

23 Leroy Hood, quoted in Joel Davis, "Leroy Hood: Automated Genetic Profiles," *Omni*, November 1987, p. 118.

24 Martha Volner, quoted in Geoffrey Cowley, "Flunk the Gene Test and Lose Your Insurance," *Newsweek*, December 23, 1996, p. 48.

25 Nancy Wexler, quoted in Lauren Picker, "All in the Family," *American Health*, March 1994, p. 24.

26 Francis Galton, quoted in William Cookson, *The Gene Hunters* (London: Aurum Press, 1994), p. 24.

27 *Buck v. Bell*, 274 U.S. 200.

28 *Skinner v. Oklahoma*, 316 U.S. 535.

29 Dean Rosen, quoted in Christopher Hallowell, "Playing the Odds," *Time*, January 11, 1999, p. 60.

30 Anonymous insurance executive, quoted in Hallowell, p. 60.

31 Sandra Day O'Connor, quoted in *Washington Post*, "Disability Act Doesn't Cover Correctable Ailments," reprinted in *San Francisco Chronicle*, June 23, 1999, p. A1.

32 Harvie Raymond, quoted in Seth Shulman, "Preventing Genetic Discrimination," *Technology Review*, July 1995, p. 17.

33 Lori B. Andrews, quoted in Rick Weiss, "Predisposition and Prejudice," *Science News*, January 21, 1989, p. 41.

34 Paul Billings, quoted in Bettyann H. Kevles and Daniel Kevles, "Scapegoat Biology," *Discover*, October 1997, p. 62.

35 W. French Anderson, "Gene Therapy," *Scientific American*, September 1995, p. 124.

36 W. French Anderson, quoted in Joseph Levine and David Suzuki, *The Secret of Life* (Boston: WGBH Educational Foundation, 1993), p. 207.

37 Bill Clinton, quoted in David Perlman and Charles Petit, "Clinton Bans Human Clone Funding." *San Francisco Chronicle*, March 10, 1997, p. A1.

38 Harold Varmus, quoted in Beth Baker, "To Clone or Not to Clone—Congress Poses the Question," *Bioscience*, June 1997, p. 340.

39 Ian Wilmut, quoted in "Cloning Raises Difficult Issues for White House, Congress," in *Issues in Science and Technology*, Summer 1997, p. 22.

40 *Lifchez v. Hartigan*, quoted in Mark D. Eibert, "Clone Wars," *Reason*, June 1998, p. 53.

CHAPTER 2

THE LAW AND BIOTECHNOLOGY

LAWS AND REGULATIONS

Hundreds of pieces of legislation, regulations, and policy statements relating to biotechnology and genetic engineering have been issued by the U.S. Congress, the congresses of the states, or other government bodies such as the National Institutes of Health (NIH) and the National Bioethics Advisory Commission. This section details some of the best known and most important ones. They are grouped according to the five chief topics discussed in Chapter 1 (Agricultural Biotechnology and Safety, Patenting Life, DNA "Fingerprinting," Genetic Health Testing and Discrimination, and Human Gene Alteration and Cloning). Within each topic, they are arranged by date.

Agricultural Biotechnology and Safety

RECOMBINANT DNA ADVISORY COMMITTEE CHARTER

In accordance with Section 402 (b) (6) of the Public Health Service Act (42 USC 282), the National Institutes of Health in October 1974 established the Recombinant DNA Advisory Committee, consisting of 15 members. At least eight of the committee members were to be experts in recombinant DNA research, molecular biology, or similar fields, and at least four were to be experts in applicable law, standards of professional conduct and practice, public attitudes, the environment, public health, occupational health, or related fields. The committee's job is to advise the NIH director concerning the current state of knowledge and technology regarding DNA recombinants and to recommend guidelines to be followed by investigators working with recombinant DNA. The director is required to consult it before making major changes in existing NIH guidelines. In the early years of its existence, the RAC had to approve most new types of genetic engineering experiments. Today, however, its approval is seldom required. The RAC's most recent charter was approved by NIH Director Harold Varmus on May 27, 1997.

NATIONAL INSTITUTES OF HEALTH GUIDELINES

Almost immediately after the development of recombinant DNA technology in the early 1970s, some scientists became concerned about the safety of experiments that, for instance, would insert cancer-causing genes into bacteria that could live in the human small intestine. That concern led to a historic conference at Asilomar, California, in February 1975, during which some 140 geneticists and molecular biologists hammered out guidelines that classified recombinant DNA experiments into four groups according to the degree of danger they represented. The scientists recommended that experiments classified as P3 or P4, the most dangerous categories, not be carried out until sufficiently secure facilities could be built.

The Asilomar recommendations, including the P1–P4 classification system, were expanded into guidelines issued by the NIH in June 1976. These were the first federal regulations governing the new technology. They dealt primarily with containment requirements for experiments in the different categories and were aimed at preventing the accidental release of genetically altered microorganisms into the environment. The regulations were binding only on laboratories receiving federal funding. Legal authority for the NIH or, for that matter, any other single federal agency to oversee recombinant DNA technology was almost nonexistent, but most private (especially university) laboratories agreed to abide by the guidelines.

Realizing that time had passed and no disasters had occurred, the NIH relaxed its guidelines in 1978 and again in 1980, downgrading many experiments to lower containment levels and exempting others—eventually, most—from all regulation. The NIH guidelines still exist in modified form, retaining the four-level biosafety classification system and specifying the containment requirements for each level. The guidelines also state which types of genetic engineering experiments require approval of the RAC, the NIH, institutional biosafety committees, or some combination of these.

They list special requirements for human gene therapy experiments. The guidelines were most recently revised in May 1999.

COORDINATED FRAMEWORK FOR REGULATION OF BIOTECHNOLOGY

In June 1986, the federal Office of Science and Technology Policy published the Coordinated Framework for Regulation of Biotechnology, which divided regulation of genetic engineering research and technology among five agencies: the NIH, the National Science Foundation, the U.S. Department of Agriculture (USDA), the Environmental Protection Agency (EPA), and the Food and Drug Administration (FDA). The first two groups were to evaluate research supported by government grants. The latter three agencies were, and are, the chief regulators of the environmental testing and sale of biotechnology products, under the authority of several laws that were amended to include genetically altered organisms. The agencies consider whether such products are safe to grow, safe for the environment, and (if intended as food or feed) safe to eat. State laws, such as seed certification laws, may also affect bioengineered products.

FEDERAL FOOD, DRUG, AND COSMETIC ACT (FFDCA)

First passed in 1938, the Federal Food, Drug, and Cosmetic Act (21 USC 9) has been amended to give both the EPA and the FDA control over certain biotechnology products. The FFDCA gives the EPA the right to set tolerance limits for substances used as pesticides on and in food and feed. This includes tolerances for residues of herbicides used on food crops genetically altered to be herbicide tolerant and for pesticides in food crops that produce such substances. The FFDCA gives the FDA the power to regulate foods and feed derived from new plant varieties, including those that are genetically engineered. It requires that genetically engineered foods meet the same safety standards required of all other foods.

The FFDCA was most recently amended on May 19, 1992. The FDA's current biotechnology policy under FFDCA treats substances intentionally added to food through genetic engineering as food additives if they are significantly different in structure, function, or amount than substances currently found in food. The agency has concluded, however, that many genetically altered food crops do not contain substances significantly different from those already in the diet and thus do not require FDA approval before marketing. This 1992 ruling has been criticized by those who believe that genetically modified food crops may threaten human health.

FEDERAL INSECTICIDE, FUNGICIDE, AND RODENTICIDE ACT (FIFRA)

The Federal Insecticide, Fungicide, and Rodenticide Act (7 USC 136) was passed in 1947 to regulate the distribution, sale, use, and testing of chemical and biological pesticides. After the EPA was established in 1970, it took over regulation of pesticides under FIFRA. FIFRA has been amended to include plants and microorganisms producing pesticidal substances, such as agricultural crops genetically modified to produce *Bacillus thuringensis* (Bt) toxin.

TOXIC SUBSTANCES CONTROL ACT (TSCA)

The Toxic Substances Control Act (15 USC 53), passed in 1976, gives the EPA the authority to, among other things, review new chemicals before they are introduced into commerce. Section 5 of TSCA was amended in 1986 to classify microorganisms intended for commercial use that contain or express new combinations of traits, including "intergeneric microorganisms," which contain combinations of genetic material from different genera, as "new chemicals" subject to EPA regulation under the act. The EPA believes that organisms containing genes from such widely separated groups are sufficiently likely to express new traits or new combinations of traits to justify being termed "new" and reviewed accordingly. The EPA handles this review under its Biotechnology Program. Altered microorganisms containing genetic material from two species in the same genus are not subject to regulation under TSCA.

EPA regulations under TSCA were amended on April 11, 1997, to tailor the general screening program for microbial products of biotechnology to meet the special requirements of microorganisms used commercially for such purposes as production of industrial enzymes and other specialty chemicals; creation of agricultural aids such as biofertilizers; and breakdown of chemical pollutants in the environment (bioremediation). According to the EPA, this change provides regulatory relief to those wishing to use these products of microbial biotechnology while still ensuring that the agency can identify and regulate risk associated with such products.

PLANT PEST ACT

The Animal and Plant Health Inspection Service (APHIS) is the agency within the USDA that is responsible for protecting United States agriculture from pests and diseases. Under the Plant Pest Act (7 USC 7B), originally passed in 1987, APHIS regulations provide procedures for obtaining a permit or for providing notification to the agency before introducing into the

United States, either by import or by release from a laboratory, any "organisms and products altered or produced through genetic engineering which are plant pests or which there is reason to believe are plant pests." The law was amended on March 31, 1993, and again on May 2, 1997, to simplify requirements and procedures. The latter amendment made notification, rather than obtaining of a permit, sufficient for the release of most new types of genetically engineered plants into the environment.

Patenting Life

PATENT LAW IN THE CONSTITUTION

Article I, Section 8, of the U.S. Constitution gives Congress the power to enact laws relating to patents—that is, to "promote the progress of science and useful arts, by securing for limited times to authors and inventors the exclusive right to their respective writings and discoveries." The first patent law was passed in 1790.

1952 PATENT LAW REVISION

Current patent law stems from Title 35 of the U.S. Code, which was revised on July 19, 1952. The law specifies the requirements for patentability and the procedure for obtaining patents. It also gives the Patent and Trademark Office the job of granting patents and administering patent regulations.

The parts of Title 35 of greatest concern to biotechnology are Sections 100 to 103, which describe the criteria that determine which inventions are patentable. Section 101 states that a patentable "process, machine, manufacture, or composition of matter, or any . . . improvement thereof" must be "new and useful." Section 102 further defines the requirement of novelty. Section 103 adds the qualification that a patent may not be obtained if "the subject matter [of the item to be patented] as a whole would have been obvious at the time the invention was made to a person having ordinary skill in the art to which said subject matter pertains." The third paragraph of Section 103, added later, specifically adds biotechnological processes, including gene alteration and production of cell lines, to the list of patentable items.

The 1952 patent law was cited in the landmark Supreme Court case *Diamond v. Chakrabarty*, which in 1980 allowed living things other than plant varieties to be patented for the first time. In his majority opinion on that case, Chief Justice Warren Burger referred to testimony accompanying the

1952 law in which Congressman P. J. Federico, a principal drafter of the legislation, stated that Congress intended it to "include anything under the sun that is made by man."

PLANT PATENT ACT

Traditionally, living things were held to be "products of nature" and thus not patentable. In 1930, however, Congress passed the Plant Patent Act (35 USC 15), which provides that "whoever invents or discovers and asexually reproduces any distinct and new variety of plant" may obtain a patent on it. The patent grants "the right to exclude others from asexually reproducing the plant or selling or using the plant so reproduced." Only asexual reproduction was mentioned in this act because, at the time, hybrids could not be made to breed true (reproduce sexually).

PLANT VARIETY PROTECTION ACT

By 1970, it had become possible to reproduce hybrid plant varieties sexually, that is, by seed. Congress therefore passed the Plant Variety Protection Act (7 USC 57), which extends the patent or patentlike protection of the Plant Protection Act to plants that could be reproduced in this way. The variety protected has to be new, distinct, uniform, and stable. Except for farmers, who are allowed to save and reuse seeds under certain circumstances, the act forbids unauthorized sexual reproduction of protected plant varieties. The act specifically exclude fungi and bacteria from its coverage.

DNA "Fingerprinting"

DNA IDENTIFICATION ACT

On September 13, 1994, Congress passed the DNA Identification Act (42 USC Sec. 14131). This act attempted to answer criticisms of the quality of forensic DNA testing raised in such court cases as *New York v. Castro* (1989) and *United States v. Yee* (1990) by ordering the FBI to establish standards for quality assurance and proficiency testing of laboratories and analysts carrying out such testing. The FBI was supposed to set up a system of blind external proficiency testing (that is, testing done by an outside agency) for forensic DNA laboratories within two years, unless the agency concluded that such a system was not feasible. The FBI, however, has continued to allow laboratories to test themselves. Critics have questioned the wisdom of giving the FBI control over DNA quality assurance, both because it has its own DNA laboratory and because investigators have found major flaws in the quality of that laboratory's testing.

Genetic Health Testing and Discrimination

AMERICANS WITH DISABILITIES ACT (ADA)

Passed in 1990, the Americans with Disabilities Act (42 USC 12101–12111, 12161, 12181) was intended to increase disabled people's access to public spaces, communication, transportation, and jobs and to prevent discrimination against them in employment and other areas. Section 12102 of the act defines disability as having to meet one of three possible criteria: (1) a physical or mental impairment that substantially limits one or more of the major life activities of an individual; (2) a record of such impairment; or (3) being regarded as having such impairment. People suffering from inherited diseases would surely be considered disabled, but it is not yet clear whether people who are presently healthy but are likely to develop an inherited illness later in life (as in late-onset diseases such as Huntington's disease) or have inherited a gene associated with increased risk of an illness such as cancer can be considered disabled. The Equal Employment Opportunities Commission ruled in 1995 that using genetic test results to deny employment to people in this category was discrimination under the ADA, but its ruling does not have the force of law.

HEALTH INSURANCE PORTABILITY AND ACCOUNTABILITY ACT

The Health Insurance Portability and Accountability Act, passed on August 21, 1996 (P.L. 104–191), was intended primarily to help people keep their health insurance when they change jobs. A paragraph in Section 701, "Increased Portability Through Limitation on Preexisting Condition Exclusions," forbids considering genetic information as a preexisting condition for insurance purposes unless a person is actually suffering from an inherited disease. Thus, for instance, a woman who is shown by a test to have a mutated form of the gene BRCA1, which is associated with an increased risk of breast and uterine cancer, but who does not actually have cancer could not be denied insurance. Over 30 states also have laws forbidding employers, insurers, or both to obtain genetic information or to use it in making decisions.

Human Gene Alteration and Cloning

THE NATIONAL BIOETHICS ADVISORY COMMISSION REPORT

On February 27, 1997, Ian Wilmut and his colleagues at the Roslin Institute in Scotland startled the world by announcing that they had cloned a sheep, Dolly, from an udder cell taken from a six-year-old Finn Dorset ewe. Although intended primarily as an advance in agricultural biotechnology, this first cloning of a mammal from a mature body cell spawned fears that humans would soon be cloned as well. About a week after Wilmut's story hit the headlines, U.S. president Bill Clinton called for a ban on the use of federal funds for research on human cloning. He also asked the National Bioethics Advisory Commission (NBAC), a group of 18 experts in theology, science, and law headed by Princeton University president Harold Shapiro, to study the issue for 90 days and provide recommendations.

The NBAC issued its report in June 1997. The report's chief conclusion was that "at this time it is morally unacceptable for anyone in the public or private sector, whether in a research or clinical setting, to attempt to create a child using somatic cell nuclear transfer cloning" (the technique used to create Dolly) because such a technique would be "likely to involve unacceptable risks to the fetus and/or potential child."

The NBAC supported Clinton's call for a federal funding moratorium on research aimed at producing a child through cloning and asked that privately funded researchers also agree to halt work on this line of research. It stressed, however, that any legislation banning human cloning research should have a "sunset clause" that would require a review of the situation in three to five years. The commission stated that any ban on creation of a child by cloning "should be carefully written so as not to interfere with other important areas of scientific research," such as the cloning of animals or of human DNA sequences and cell lines.

As of late 1999, no federal law concerning human cloning has been enacted. The NBAC's recommendations have no legal power. They are of interest, however, because they are the chief scientific/government policy statement on the subject to date.

COURT CASES

A great deal of litigation has arisen over issues related to biotechnology and genetic engineering. In biotechnology, many have been patent disputes.

In human genetics and genetic engineering, most have related to either DNA fingerprinting or genetic discrimination. This section describes some of the important court cases relating to patenting of living things (including products of the human body), forensic DNA testing, and genetics-based discrimination in insurance and employment (including eugenics laws).

Patenting Life

DIAMOND V. CHAKRABARTY, 447 U.S. 303 (1980)

Background

Ananda Chakrabarty, a scientist working for General Electric Corporation, modified a bacterium of the genus *Pseudomonas* so that it could digest crude petroleum, something no natural bacterium could do. He did so by getting the bacterium to take up four types of plasmids from other bacteria that could digest different components of crude oil, a technique that could be classified as genetic engineering but did not involve recombinant DNA (the individual plasmids were unaltered).

In June 1972, believing that his new bacterium would be useful in cleaning up oil spills, Chakrabarty applied for patents (in the name of General Electric) on the process of making the bacteria, a mixture in which they could be spread on water, and the bacteria themselves. The Patent and Trademark Office (PTO) granted the first two patent requests but rejected the third on the grounds that, as living things, the bacteria were "products of nature" rather than "manufactures" and thus not patentable under United States law. The Patent Office Board of Appeals affirmed the PTO's decision. The U.S. Court of Customs and Patent Appeal (CCPA), however, reversed it.

The government appealed the decision to the Supreme Court in 1979. At first the court refused to hear the case, telling the CCPA to reconsider its decision in light of the comment that the court had made in a 1978 case, *Parker v. Flook*, in which the judges had said they believed they should "proceed cautiously when we are asked to extend patent rights into areas wholly unforeseen by Congress." The CCPA held its ground, however, and the court accepted the case later in 1979. It was argued on March 17, 1980, under the full name *Sidney A. Diamond, Commissioner of Patents and Trademarks, v. Ananda M. Chakrabarty*.

The Law and Biotechnology

Legal Issues

The question before the Supreme Court was technically a narrow one: interpreting the language of 35 U.S.C. 101 (part of the revision of patent law that Congress had passed in 1952) to determine "whether a live, human-made micro-organism is patentable subject matter"—that is, whether it was a "manufacture" or "composition of matter" as Congress intended the 1952 law to be construed. The root of the question, however, was deeply significant: Could a living thing—even one whose genes had been deliberately altered—be considered to be a human invention?

Traditionally, laws of nature, physical phenomena ("products of nature"), and abstract ideas have been held not to be patentable. In testimony reports accompanying the 1952 law, however, Congressman P. J. Federico, a principal drafter of the legislation, stated that Congress intended it to "include anything under the sun that is made by man."

Congress had already allowed the patenting, or at least a patentlike protection, of plant varieties in the Plant Patent Act of 1930 (which permitted plant breeders to keep exclusive rights to asexual reproduction of new varieties they developed) and the Plant Variety Protection Act of 1970 (which extended the protection to sexual reproduction of new plant varieties). House and Senate reports accompanying the 1930 act said that its purpose was to "remove the existing discrimination between plant developers and industrial inventors" because the acts of developing new plant varieties and new compositions of nonliving matter were conceptually equivalent.

The CCPA, in defending its decision, claimed that Chakrabarty's altered bacterium met the patent criteria of novelty, usefulness, and nonobviousness specified in the 1952 law (35 U.S.C. 101-103).

We look at the facts and see things that do not exist in nature and that are man-made, clearly fitting into the plain terms "manufacture" and "compositions of matter." We look at the statute and it appears to include them. We look at legislative history and we are confirmed in that belief. We consider what the patent statutes are intended to accomplish and the constitutional authorization, and it appears to us that protecting these inventions, in the form claimed, by patents will promote progress in very useful arts.

The CCPA further maintained that the fact that the things that had been "manufactured"—that is, deliberately altered by humans—were alive made no difference for patenting purposes: "[There is] no *legally* significant difference between active chemicals which are classified as 'dead' or organisms used for their *chemical* reactions which take place because they are 'alive' [emphasis in original]. Counsel for the PTO argued, on the other hand, that

Congress did not want bacteria to be patentable because the Plant Variety Protection Act specifically excluded them. In general, the PTO said, Congress had not resolved the question of whether living things should be considered patentable.

Amicus curiae ("friend of the court") briefs filed by various groups brought up several other issues. Biotechnology critic Jeremy Rifkin's organization, the People's Business Commission (later the Foundation on Economic Trends), for instance, filed a brief urging that Chakrabarty's patent be refused because of the possible threats genetically altered organisms posed to human health and the environment. The brief claimed that granting the Chakrabarty patent would provide an incentive for commercial exploitation of the new gene-altering technology that was not in the public interest. The biotechnology company Genentech, conversely, filed a brief calling for acceptance of the patent, pointing to the 1978 relaxation of the NIH guidelines as evidence that most scientists no longer feared recombinant DNA technology.

Other briefs questioned the ethics of "owning" organisms or claiming to have made them, saying that only God could make living things. Representatives of the rapidly growing biotechnology industry, on the other hand, stressed that patent protection was vital to the industry's growth. They said that if patents on engineered organisms were not allowed, much of the knowledge being generated in the new field would remain hidden in the form of trade secrets rather than being revealed so that others could use it once the patents expired.

Decision

On June 16, 1980, the Supreme Court decided by a 5-4 vote that "a live, human-made micro-organism is patentable subject matter under [U.S.C.] 101. Respondent's micro-organism constitutes a 'manufacture' or 'composition of matter' within that statute." Chief Justice Warren Burger, writing the court's majority opinion, stated,

> *The patentee has produced a new bacterium with markedly different characteristics from any found in nature and one having the potential for significant utility. His discovery is not nature's handiwork, but his own; accordingly it is patentable subject matter.*

The court concluded that Congress had intended the patent laws to have wide scope. Burger rejected the patent office's argument that the fact that Congress had passed the two plant patent protection acts meant that it had not intended the original 1952 statute to cover living things. He cited congressional commentary on the 1930 act stating that the work of the plant breeder "in aid of nature" was patentable invention. The distinction, he said, was not between living and nonliving things but between unaltered products of nature and things that had been invented by humans. The exclusion of bacteria from the 1970 act, Burger said, meant only that bacteria were not considered to be plants, not that they were not considered to be patentable.

The patent office's second argument against Chakrabarty's patent was that microorganisms could not be patented until Congress expressly authorized such protection because Congress had not foreseen genetic engineering technology at the time it passed the 1952 law. Somewhat reversing the court's cautious position in *Parker v. Flook*, Burger wrote that he "perceive[d] no ambiguity" in Congress's language that would exclude genetically altered organisms.

Justice William Brennan wrote the minority opinion. He claimed that in the plant patent acts, Congress chose "carefully limited language granting protection to some kinds of discoveries, but specifically excluding others," including bacteria. "If newly developed living organisms not naturally occurring had been patentable under 101, the plants included in the scope of the 1930 and 1970 Acts could have been patented without new legislation," he stated. In conclusion, recalling the recommendation for caution that the court had expressed in *Flook*, Brennan commented,

> *I should think the necessity for caution is that much greater when we are asked to extend patent rights into areas Congress has foreseen and considered but not resolved. . . . [The majority's decision] extends the patent system to cover living material even though Congress plainly has legislated in the belief that [present law] does not encompass living organisms. It is the role of Congress, not this Court, to broaden or narrow the reach of the patent laws.*

Finally, the question of the dangers of genetically engineered organisms—the description of which Chief Justice Burger called "a gruesome parade of horribles"—was an issue beyond the court's competence, Burger concluded. "Arguments against patentability . . . , based on potential hazards that may be generated by genetic research, should be addressed to the Congress and the Executive, not to the Judiciary," he wrote. He also noted:

> *The grant or denial of patents on micro-organisms is not likely to put an end to genetic research or to its attendant risks. The large amount of research that has already occurred when no researcher had sure knowledge that patent protection would be available suggests that legislative or judicial fiat as to patentability will not deter the scientific mind from probing into the unknown any more than Canute could command the tides.*

Impact

Commentators such as patent lawyer Mitchel Zoler have claimed that from a strict legal standpoint, the *Chakrabarty* decision was "trivial law."[1] It broke no new legal ground, but rather provided only a minor clarification of existing patent laws. Furthermore, Donald Dunner, another patent lawyer, noted in the year following the decision that the ruling was "important but . . . not life or death of the [biotechnology] industry, and even had it gone the other way, it probably would not have been."[2] Even if altered microorganisms themselves had not been considered patentable, the processes of making them would have been, or their nature could have been kept hidden as trade secrets. Furthermore, the ease of altering bacteria suggested that patents, even once obtained, could be fairly easily circumvented by further alteration.

The psychological impact of this Supreme Court decision on both supporters and critics of biotechnology, however, was enormous. By the time the court made its ruling, dozens of patent applications for recombinant and other genetically engineered organisms, mostly bacteria, had been submitted to the PTO but had not been ruled upon. The PTO now felt free to begin granting these patents. The court's decision, which was widely publicized, gave a considerable boost to the biotechnology industry and encouraged those who were thinking of investing in it. Peter Farley of Cetus Corporation, a leading biotechnology business that had been among those filing *amicus* briefs in the case, said afterward that "the positive impact [of the decision] was in the Court's bringing genetic engineering as a commercial enterprise to the attention of the entire country."[3] The decision also ignited public discussion about whether patenting living things was ethical and to what degree a patent implied "ownership" of a life-form.

The PTO expanded the *Chakrabarty* ruling in 1987, extending patent protection to animals, cells and cell lines, body parts, plasmids, and genes, including human ones. After that, the only type of living thing that could not be patented was a whole human being.

JOHN MOORE V. REGENTS OF CALIFORNIA, 51 CAL. 3D 120 (1990)

Background

John Moore, a Seattle businessman, learned in 1976 that he suffered from a rare form of leukemia. He went to the Medical Center of the University of California, Los Angeles (UCLA), for treatment, where his case was assigned to David W. Golde. Informing Moore "that he had reason to fear for his life," Moore later alleged, Golde recommended removing Moore's enlarged spleen (an abdominal organ that makes blood cells) as a treatment for the disease. Moore consented, and his spleen was removed on October 20.

Moore alleged later that, unknown to him, Golde had noticed even before the operation that Moore's blood cells had an extraordinary ability to make certain immune system chemicals called lymphokines, which have potential commercial use as medical treatments because they stimulate production of immune cells that fight bacterial infections and cancer. Golde therefore "formed the intent and made arrangements to obtain portions of [Moore's] spleen following its removal." Furthermore, Golde called Moore back from Seattle for several additional treatments between 1976 and 1983, during which he took samples of Moore's blood, bone marrow, sperm, and skin.

During this same period, Golde worked with Shirley Quan, another UCLA researcher, to develop Moore's cells into an immortal cell line that was eventually named Mo after Moore. The two researchers and their employers, the Regents of the University of California, applied for a patent on their cell line in 1981, and the patent was granted in 1984. They then made lucrative contracts with Genetics Institute and the drug company Sandoz for use of the cell line.

At no time did Golde or anyone else involved in the research tell Moore about the cell line or the patent; indeed, Moore alleged, Golde repeatedly denied any commercial plans when Moore directly asked him about such a possibility. Moore nonetheless somehow found out about the extremely lucrative use to which his cells were being put (one estimate placed the potential value of products from the patented line at $3 billion). He filed suit in 1984, naming Golde, Quan, the Regents, and the drug companies as defendants. He claimed that he still "owned" his removed cells, at least in the sense that he had a right to say what use was made of them, and that he had a proprietary interest in any products that these or other researchers created with the cell line made from the cells.

A California superior court denied Moore's right to sue in 1986, but he appealed the decision, and an appellate court reversed it two years later, stating that "the essence of a property interest—the ultimate right of control—. . . exists with regard to one's human body." Golde, the university, and the other defendants appealed the case to the California Supreme Court, which heard it in July 1990.

Legal Issues

The California Supreme Court agreed to rule on whether John Moore had grounds to sue the defendants for using his cells in potentially lucrative research without his permission. The basic question was whether a person has an ownership right to body cells and tissues after they have been removed and, if so, whether this right entitles the person to compensation if others develop those cells or tissues into a commercial product.

"In effect," wrote Judge Panelli in the court's majority opinion, "what Moore is asking us to do is to impose a tort duty on scientists to investigate the consensual pedigree of each human cell sample used in research"—something that no court had ruled on before. This was an important issue. An Office of Technology Assessment report to Congress had noted in 1987,

> *Uncertainty about how courts will resolve disputes between specimen sources and specimen users could be detrimental to both academic researchers and the infant biotechnology industry, particularly when the rights are asserted long after the specimen was obtained. . . . The uncertainty could affect product developments as well as research. Since inventions containing human tissues and cells may be patented and licensed for commercial use, companies are unlikely to invest heavily in developing, manufacturing, or marketing a product when uncertainty about clear title exists. . . . Resolving the current uncertainty may be more important to the future of biotechnology than resolving it in any particular way.[4]*

The case also raised questions about what information a doctor has a "fiduciary responsibility" to give a patient in order to obtain truly informed consent before doing a medical procedure. To a lesser extent, it considered what financial or other responsibility secondary parties that have an interest in research done on a person's tissues but have no direct dealings with the person, such as a university or a drug company, bear to the person from whom the tissues came. Finally, as Judge Arabian wrote in a separate opinion concurring with the majority, the Moore case raised "the moral issue" of whether there is "a right to sell one's own body tissue for profit."

Decision

The California Supreme Court decided by a 5-2 vote that John Moore did not have a right of ownership over his tissues or cells once they had been removed from his body. "Moore's novel claim to own the biological materials at issue in this case is problematic, at best," Judge Panelli wrote in his majority opinion. Moore therefore could not sue under tort law for a conversion—essentially, a theft—of those parts. The court defined conversion as "a tort that protects against interference with possessory and ownership interests in personal property" and concluded that "the use of excised human cells in medical research does not amount to a conversion." Moore also had no direct right to income from the cell line patent because it represented invention on the part of the researchers, not any creative effort by himself, the court ruled.

In his majority opinion, Judge Panelli also expressed concern that allowing Moore to sue on the grounds of conversion

> . . . *would affect medical research of importance to all of society. . . . The extension of conversion law into this area [use of cells after they have legitimately been removed from the body] will hinder research by restricting access to the necessary raw materials. . . . Th[e] exchange of scientific materials [cell lines] . . . will surely be compromised if each cell sample becomes the potential subject matter of a lawsuit.*

Furthermore, Panelli wrote, "The theory of liability that Moore urges us to endorse threatens to destroy the economic incentive to conduct important medical research" because of the danger of lawsuits from disgruntled patients whose cells had been transformed into cell lines.

Although it rejected Moore's right to sue for conversion, the court found that David Golde had violated his "fiduciary duty" to Moore by not telling him about the proposed for-profit use of his cells. Golde's commercial plans represented a potential conflict of interest with his role as Moore's physician, and his failure to inform Moore of those plans denied Moore some of the facts he needed in order to give informed consent to the spleen operation. Judge Panelli wrote:

> *We hold that a physician who is seeking a patient's consent for a medical procedure must, in order to satisfy his fiduciary duty and to obtain the patient's informed consent, disclose personal interests unrelated to the patient's health, whether research or economic, that may affect his medical judgment.*

The court therefore ruled that Moore could sue Golde, at least, on the grounds that Golde had violated his fiduciary duty and that he had failed to obtain Moore's properly informed consent. Grounds for suing the other defendants were more dubious, though the court did not rule out the possibility of such a suit.

All the judges agreed that Golde had violated his duties to Moore, but they had differing opinions about the larger question of Moore's ownership of his body tissues. In a separate concurring opinion, Judge Arabian wrote about "the moral issue" involved in the case. He claimed that allowing Moore to sue for conversion—in other words, support Moore's claim that he owned his tissues after their removal and should have been paid for them— would be to "recognize and enforce a right to sell one's own body tissue for profit," an idea of which Arabian clearly did not approve. The judge wrote:

[Moore] entreats us to regard the human vessel—the single most venerated and protected subject in any civilized society—as equal with the basest commercial commodity. He urges us to commingle the sacred with the profane. He asks much.

Arabian expressed fears that supporting Moore's claim would result in "a marketplace in human body parts" and stated that the state legislature should settle the question of whether such a situation was permissible.

Judges Broussard and Mosk expressed other viewpoints in separate dissenting opinions. Broussard supported Moore's right to sue for conversion because of the allegation that Golde had been planning his research before suggesting that Moore have his spleen removed. By not telling Moore about his plans, therefore, Golde had interfered with Moore's ownership rights to his cells *before* the cells had been removed. Broussard pointed out that the state's Uniform Anatomical Gift Act allowed people to specify donation of their organs for transplantation after their death and claimed that this fact supported the idea that people could say how they wanted donated parts of their body to be used. Broussard wrote:

The act clearly recognizes that it is the donor of the body part, rather than the hospital or physician who receives the part, who has the authority to designate, within the parameters of the statutorily authorized uses, the particular use to which the part may be put.

Unlike Panelli, Broussard did not feel that allowing occasional suits like Moore's (which had the unusual feature that the commercial usefulness of his cells had been discovered, and actively concealed from Moore, before the cells had been removed from his body) would put a damper on medical research with cell lines or that, even if it did, this was sufficient reason to deny Moore's right to sue. Because most of the value of the cell line patent lay in the researchers' work, Broussard suspected that the damages Moore would receive would be relatively small even if he won his suit.

Broussard's view of the effect the court's decision would have on the possible sale of cells or body parts was exactly the opposite of Arabian's. Broussard wrote:

Far from elevating these biological materials above the marketplace, the majority's holding simply bars plaintiff, the source of the cells, from obtaining the benefit of the cells' value, but permits defendants, who allegedly obtained the cells from plaintiff by improper means, to retain and exploit the full economic value of their ill-gotten gains free of their ordinary common law liability for conversion.

Judge Mosk also dissented from some of the majority's decisions and reasonings. If past judicial rulings did not cover ownership of body parts, Mosk saw no reason not to extend them:

> *If the cause of action for conversion is otherwise an appropriate remedy on these facts, we should not refrain from fashioning it simply because another court has not yet so held or because the Legislature has not yet addressed the question.*

Mosk also felt that, although Moore had contributed no creative effort toward development of the patented cell line made from his spleen, he was nonetheless a kind of "joint inventor."

> *What . . . patients [like Moore] . . . do, knowingly or unknowingly, is collaborate with the researchers by donating their body tissue. . . . By providing the researchers with unique raw materials, without which the resulting product could not exist, the donors become necessary contributors to the product.*

Because of that contribution, Mosk said, Moore should be entitled to some compensation.

Mosk agreed with Broussard that a threat to medical research on cell lines was not a sufficient reason to deny Moore's claim, though he gave different reasons for his view. Secrecy and competition in the biotechnology industry had already severely inhibited the exchange of information and research materials, Mosk wrote. Furthermore, he claimed, researchers would know where their cells came from and whether proper consent for their use had been obtained if they engaged in "appropriate recordkeeping."

Above all, Mosk, like Broussard, felt that denying Moore's right to own his body parts would result in the human body being treated as a salable product and thus was morally reprehensible. Mosk quoted an earlier judicial decision that stated:

> *The dignity and sanctity with which we regard the human whole, body as well as mind and soul, are absent when we allow researchers to further their own interests without the patient's participation by using a patient's cells as the basis for a marketable product.*

Impact

John Moore eventually filed suit on the grounds that the court had left open. His suit was settled out of court.

The results of the Moore trial pleased biotechnology researchers, who had feared possible liability from working with cell lines or having to share revenue from them if Moore's right to sue was upheld. It disappointed those who disapproved of the patenting of living things or of tissues, cells, or genes taken from human beings. At the same time, it discouraged the establishment of a marketplace or "body shop" where human organs, tissues, or cells would be bought and sold; at least, they would not be sold by their original owners or while still residing in those owners' bodies.

Although the Moore case is often cited as establishing the principle that people do not own rights to their body tissues, the ruling does not apply outside California (or even necessarily to all cases within California). Some other states have different laws. In Oregon, for instance, a 1995 law (amended in 1997) specifically grants ownership rights over tissues and the genetic information derived from them to the people from whose bodies the tissues came. There have been no national rulings on this issue.

DNA "FINGERPRINTING"

FLORIDA V. ANDREWS, 533 SO. 2D 841 (FLORIDA 5TH DISTRICT COURT OF APPEALS, 1988)

Background

In 1986, a number of women were raped, beaten, and cut in Orlando, Florida. The intruder entered their homes late at night, when they were alone, and evidence suggested that he had stalked them before the attacks to learn their habits. He covered each woman's head with a blanket or sheet, so only one of his victims, Nancy Hodge, saw his face. Police, however, were able to obtain semen samples from Hodge and one other woman, a young mother.

Responding to a report of a prowler in early 1987, police captured Tommie Lee Andrews, a 24-year-old warehouse worker. Nancy Hodge picked out Andrews's picture from a photo lineup, and he was charged with her rape and that of the other woman who had provided a semen sample.

Tim Berry, Andrews's prosecutor, was reluctant to base his case entirely on Hodge's identification. A blood typing test suggested that Andrews could have committed the rapes—but so could 30 percent of the men in the United States. That summer, however, another attorney told Berry about the DNA "fingerprinting" technique that British geneticist Alec Jeffreys had invented a few years before. The technique had been used in numerous immigration and paternity cases and had just made its first appearance in a British criminal court. The technique allowed DNA from tiny samples of blood, semen, or other body fluids found at a crime scene to be compared with that from a suspect's blood at certain locations in the DNA molecule that differed considerably from person to person.

Berry sent the semen samples and a little of both Andrews's and Hodge's blood to Lifecodes in Valhalla, New York, one of the few laboratories in the United States then able to perform the test. The DNA in the semen samples matched that of Andrews but not Hodge. The Lifecodes analyst said the odds of the match occurring by chance (that is, of Andrews being innocent, yet still having DNA that matched that in the semen sample) were one in ten billion—almost twice the population of the world.

Legal Issues

At the time of Andrews's trial, DNA profiling evidence had been used in only one other criminal case in the United States, *Pennsylvania v. Pestinikas*, and this macabre case had not been a close parallel to that of Andrews. In 1986, Helen and Walter Pestinikas, owners of a nursing home, had allegedly allowed one of their elderly patients to starve to death and then, to avoid detection, exchanged his body's internal organs with that of a healthier deceased patient so that signs of starvation would not be found when the organs were examined. Police exhumed the man's body and compared the DNA in the organs with that elsewhere in the body. Because the DNA matched, the Pestinikases were acquitted of the organ-switching charge, though they were later convicted of allowing the man to starve.

The use of DNA profiling in the British case was much more similar to that in the Andrews situation. Jeffreys's test had shown a match between the DNA in semen found on two teenaged girls who had been raped and murdered in Leicestershire and that in the blood of Colin Pitchfork, a 27-year-old baker. During the hunt for the killer, the police had taken the unusual step of asking all men between ages 13 and 30 in several villages—some 5,000 people—to voluntarily give samples of their blood for testing. Fearing detection, Pitchfork had persuaded a coworker to give blood in his stead, but the man bragged about it and was overheard. When questioned, he led the police to Pitchfork. Pitchfork confessed and was convicted of the crimes. DNA testing had also exonerated another suspect in the case, Rodney Buckland. Buckland had confessed to one of the killings, but his DNA did not match the semen sample, so he could not have been guilty.

At the time of the Andrews trial, then, DNA "fingerprinting" had had one spectacular and widely publicized success—the Pitchfork case—and that success had occurred in a situation much like the one in which Andrews was involved. The test was still extremely new to forensics, however, and there was considerable question about whether it would meet the "*Frye* rule" (based on a 1923 case, *Frye v. United States*) by which many judges decided whether evidence from a new scientific technique would be admitted in a trial. In the *Frye* case, the court had ruled that a technique had to be "sufficiently established to have gained general acceptance in the particular field in which it belongs" before evidence based on it could be used.

Decision

Andrews's trial for the rape of Nancy Hodge took place in October 1987. In a pretrial hearing on October 19, Berry brought in an expert witness who testified that the technique on which Jeffreys's test was based, although new in the courtroom, was widely accepted in genetics and molecular biology laboratories. The judge agreed on this basis that DNA profiling met the Frye requirement, and he allowed the DNA evidence to be presented in Andrews's trial. When a Lifecodes expert brought up the one-in-ten-billion statistic, however, the defense lawyers objected. Unprepared for the challenge, he withdrew the statistic. Without it, even the DNA evidence combined with Hodge's identification of Andrews apparently was not enough. The jury was unable to reach a verdict, and the judge declared a mistrial.

Andrews went on trial for the young mother's rape a few weeks later, however, and this time the prosecutors were able to provide legal backing for the use of the statistics that supported the DNA test results. In this case, furthermore, Andrews had left literal as well as genetic fingerprints behind. He was found guilty on November 6, becoming the first person in the United States to be convicted of a crime partly on the basis of DNA evidence. In addition, he was retried for Hodge's rape in February 1988 and this time, despite questions raised by the defense lawyers, he was convicted. His total sentences for the two convictions amounted to 100 years in prison.

Impact

Coming soon after the widely publicized success of DNA profiling in the Pitchfork case, the technique's usefulness in convicting Andrews caused prosecutors to turn to it eagerly in similar cases. It was hailed as "a prosecutor's dream," the greatest aid to identifying criminals since the development of fingerprinting a century before. Judges and juries began to accept it as well. Nonetheless, its validity usually had to be established in a separate Frye hearing for each case.

NEW YORK V. CASTRO, 144 MISC. 2D 956, 545 NYS 22D 985 (1989)

Background

In February 1987, Vilma Ponce and her two-year-old daughter were stabbed to death in their apartment in the Bronx, in New York City. Ponce's common-law husband cast suspicion on the building's janitor, José Castro. Police found a spot of blood on Castro's wristwatch and had its DNA compared with that in the murder victims' blood by Lifecodes, the same laboratory that had analyzed the samples in the Tommie Lee Andrews case. In a report to the Bronx district attorney's office in July 1987, Lifecodes claimed that the watch sample matched that in Ponce's blood with only a one in 100 million chance of error. Castro was arrested and charged with the murders.

Legal Issues

In the year and a half or so that passed between the Tommie Lee Andrews trial and the beginning of José Castro's trial in February 1989, evidence from DNA "fingerprinting" tests had been used in about a hundred other trials. Few juries had been able to resist its impressiveness. As one juror was overheard to remark, "You can't argue with science."[5] Indeed, DNA testing was so impressive that it had kept hundreds of other cases from coming to trial at all: Faced with a supposedly infallible match, many defendants simply pled guilty.

In the *Castro* case, however, defense lawyers Peter Neufeld and Barry Scheck decided that they *would* argue with science. "[We] had recently become concerned about the use of DNA typing evidence in the courts," Neufeld later told *Science* magazine. "We therefore decided to make the Castro case the first in which there would be a comprehensive inquiry into the various issues that comprise DNA typing."[6]

Scheck and Neufeld did not question the basic genetic principles on which DNA profiling was based. Instead, they focused primarily on the way Lifecodes had determined a match between the two blood samples. A computer was supposed to be used in such determinations, but, when questioned by the lawyers in a pretrial Frye hearing, a Lifecodes representative admitted that the matching had been done by eye alone. The matcher had ignored two faint bands that appeared on the autoradiograph from the watch sample but not on the one from Ponce's blood. The Lifecodes expert claimed that these bands were caused by bacterial contamination, but he could present no proof of his contention. By pointing out such sources of error, Scheck and Neufeld not only cast doubt on the specific evidence in the *Castro* case but in effect, as *Science* reporter Roger Lewin said, "put forensic DNA fingerprinting as a whole on the witness stand."[7] The judge's reaction to their presentation, therefore, could have a major effect on whether DNA evidence would continue to be accepted in court.

Decision

Both the prosecution and the defense called in DNA experts to testify during the *Castro* Frye hearing. Richard Roberts of Cold Spring Harbor Laboratory on Long Island led the prosecution experts, and Eric Lander of the Whitehead Institute in Cambridge, Massachusetts, headed those who appeared for the defense. When they looked at the Lifecodes evidence and heard testimony from the laboratory's spokesman, both Roberts and Lander became dismayed. Roberts suggested that the experts from both sides meet outside the courtroom to review the situation. "We wanted to . . . look at the evidence as scientists, not as adversaries," Roberts said later.[8] The meeting, which Roberts, Lander, and two other scientists attended, took place in a New York law office on May 11, 1989.

"We all agreed that the evidence was seriously flawed," Lander wrote later.[9] The scientists composed an unprecedented joint statement to that effect and gave it to Judge Gerald Scheindlin. The statement said,

> . . . the DNA data in this case are not scientifically reliable enough to support the assertion that the samples . . . do or do not match. . . . If these data were submitted to a peer-reviewed journal in support of a conclusion, they would not be accepted.[10]

Faced with such unanimity, Judge Scheindlin, not surprisingly, ruled on August 18 that, although he considered DNA profiling evidence acceptable in the abstract, the particular evidence in this case was legally inadmissible. In spite of this, José Castro later confessed to the crime in return for a reduced sentence. He admitted that the blood on his watch had come from Ponce.

Impact

The *Castro* trial marked the first time that DNA evidence had been seriously questioned in court. "The Castro hearing put DNA typing as a whole on trial," Peter Neufeld said afterward, "and from the evidence we've seen, you'd have to say it fails."[11] The case brought up what has been, and probably still is, the technique's chief weakness: the accuracy with which the testing is carried out and interpreted. (A related problem, careless police handling of samples before testing, played a major role in the famous O. J. Simpson murder trial in 1995.) Although some of the mistakes in the Lifecodes technique, including determination of matches by eye alone, were quickly remedied, the accuracy of all the country's main forensic DNA testing laboratories has subsequently been questioned at one time or another. As Lander wrote in an article in *Nature* that described the *Castro* case, "Clinical laboratories must meet higher standards to be allowed to diagnose strep throat than forensic labs must meet to put a defendant on death row."[12]

The accuracy problem has been made worse because there is no national agency that oversees a system of outside proficiency testing for DNA laboratories. Since passage of the DNA Identification Act in 1994, the FBI has been in charge of setting standards for such laboratories. The FBI standards rely on self-testing, however, and legal and scientific experts have questioned their validity because the FBI's own crime laboratory has been shown to have what Deputy District Attorney Jamie Gorelick admitted in 1997 was "a serious set of problems" with accuracy.[13]

In the wake of *Castro*, some forensic DNA experts feared that judges would stop admitting DNA test results in future trials or that verdicts already rendered on the basis of such testing would come under question, resulting in retrials and even, perhaps, the freeing of guilty people. On the other hand, they hoped that it would encourage a drive for national standards and proficiency testing for testing laboratories. "The Castro case is likely to have a ripple effect in the legal community," Edward Imwinkelried, professor of law at the University of California, Davis, said at the time.[14] On the other hand, Roberts noted, "There is a great deal of danger of making too much of the Castro case. . . . I don't think it is typical."[15] In fact, neither hopes nor fears fully materialized. No outside proficiency testing system was established, but neither was DNA testing widely rejected.

UNITED STATES V. YEE, 134 FRD 161 (N.D. OHIO, 1991)

Background

Three Hell's Angels—Mark Verdi, John Ray Bonds, and Stephen Wayne Yee—were accused of shooting to death a record store clerk, whom they allegedly mistook for the leader of a rival gang, in his van in Sandusky, Ohio. The police found some blood in the van that did not appear to belong to the victim and concluded that one of the killers must have hurt himself during the attack and left some of his blood behind. DNA testing matched the blood to a sample from Bonds. The trio's defense lawyers called in Peter Neufeld and Barry Scheck, well known after their work on the *Castro* case, to try to keep the DNA evidence out of the courtroom by persuading the judge that the technology did not meet the Frye standard.

Legal Issues

As they had in the *Castro* case, Scheck and Neufeld put DNA profiling "on the witness stand" during *Yee*'s pretrial Frye hearing, which was held in the summer of 1990. This time, however, although they again pointed up sloppy testing and matching procedures (this time in the FBI's own laboratory), they focused chiefly on a debate among population geneticists about the assumptions underlying the method used to calculate the statistics that showed the probability of a chance match occurring between DNA samples of two different people (other than identical twins) within a large population—in other words, the chance that a suspect might be innocent and still have a DNA sample that matched one from the crime scene. Those statistics, which usually showed the chance of a false match to be one in millions or billions (sometimes more than the entire world population), impressed juries, but according to population geneticists Richard Lewontin and Daniel Hartl, they were badly flawed. (Scheck and Neufeld had also raised this issue in *Castro*, but they had not stressed it as they did in *Yee*.)

Testifying for the defense, Hartl and Lewontin explained to the judge that the statistics were calculated by the so-called multiplication rule. In this rule, the frequency of the allele, or particular DNA variation, found at each spot, or locus, examined in the test was determined for a large population, such as Caucasians, Blacks, or Hispanics. These figures were multiplied together to form the denominator of a fraction. Thus, if four loci were tested and one out of a hundred people in the appropriate large population had the particular allele found at each spot, the chance of two different people matching at all four spots would be one in 100,000,000 (100 x 100 x 100 x 100).

The problem with the multiplication rule, the experts said, was that it was based on the assumption that all the people within the large population group intermarried randomly and therefore that all the alleles were inherited independently of each other. Lewontin and Hartl questioned this assumption. They pointed out that, for instance, the population of "Hispanics" included people of Mexican ancestry, Puerto Rican ancestry, Colombian ancestry, and many more. People in each of these subpopulations tended to live near each other and marry each other. The frequency of a particular allele might thus be quite different in different subpopulations, and figures calculated from the general population statistics would probably be inaccurate for a particular group. The two geneticists recommended that some other method of calculation be used until figures for allele frequencies could be obtained for a variety of subpopulations, a task that they admitted might take 10 to 15 years.

Two equally well-regarded experts, Kenneth Kidd and Thomas Caskey, testified for the prosecution. They agreed that subpopulations existed within the larger groups and that people might be more likely to marry within their own subgroup than outside it. Caskey and Kidd did not believe that these facts had a major effect on the independent inheritance of the alleles, however. They maintained that the multiplication rule could safely be used.

Decision

In spite of the conflicting expert testimony and what he termed the "remarkably poor quality of the FBI's work and infidelity to important scientific principles," Magistrate James Carr ruled in October 1990 that the DNA evidence could be admitted in court, and the U.S. District Court for the Northern District of Ohio accepted his recommendation.[16] Yee and his codefendants were convicted of federal weapons violations and later tried for murder.

Impact

In contrast to the amicable unanimity that developed among the supposedly competing experts in *Castro*, the *Yee* hearing was an extremely adversarial one. Both prosecution and defense lawyers attacked not only the opposing scientists' professional conclusions but their personal integrity, accusing them of being influenced by high fees or financial ties to laboratories that did DNA testing.

Probably because of personal bad feeling as well as genuine scientific disagreement, the population genetics furor moved from the courtroom to the pages of *Science*, the prestigious magazine of the American Association for the Advancement of Science. Two opposing articles on the subject, one by Lewontin and Hartl and the other by Kidd and University of Texas geneticist Ranajit Chakraborty, appeared in the magazine's December 20, 1991, issue. An accompanying news article described Lewontin and Hartl's charges that they and *Science* editor Daniel Koshland had been pressured by the Department of Justice not to publish their article, a charge Koshland denied. The National Academy of Sciences' National Research Council tried to soothe hurt feelings and provide a way out of the dilemma by offering a different method of calculation in their 1992 report on forensic DNA testing, but many population geneticists questioned their method as well.

As time has passed and inheritance figures for more subpopulations have been obtained, the statistics used with DNA testing have been refined. On the whole, the results have tended to bear out the contention of Kidd, Caskey, and others that, important as subpopulations may be from a social point of view, they have relatively little effect on the relevant inheritance statistics. The statistical issue has also tended to "recede into the background," as even Hartl admitted in 1994, because DNA tests used in the late 1990s both use a more accurate method and examine more loci (usually 13), thus making the chance of accidental duplication vanishingly small in any case except, perhaps, where a test had to distinguish among several members of the same family.[17]

Genetic Health Testing and Discrimination

BUCK V. BELL, 274 U.S. 200 (1927)

Background

In 1924, when she was 18 years old, Carrie Buck, a "feebleminded" (developmentally disabled) woman in the State Colony for Epileptics and Feeble Minded in Virginia, was ordered to be sexually sterilized under a newly passed state law that required such treatment for people living in state-supported institutions who were found to have hereditary forms of insanity or subnormal intelligence. Such sterilization was supposedly necessary to promote "the health of the patient and the welfare of society." Buck was deemed to be hereditarily feebleminded because her mother was of subnormal intelligence (she was confined in the same institution) and there were signs that Buck's illegitimate baby daughter was as well.

The Virginia law was typical of laws then existing, or soon to exist, in 34 states of the United States and several other countries, including Canada (some provinces), Britain, Germany, and the Scandinavian countries. These laws were based on the "scientific" doctrine of eugenics, which had been established in the late 19th century by British scientist Francis Galton and was widely accepted at the time of the Buck case. Galton and his followers believed that complex personality traits such as intelligence were inherited, and they claimed that the human race would be improved if groups such as the subnormally intelligent, habitual criminals, and the insane were prevented from reproducing—by force, if necessary. Such action, they said, would also save society considerable money by reducing the number of people who must be incarcerated and cared for at state expense.

Buck (or others acting on her behalf) sued the director of the institution to prevent her operation. The Virginia Supreme Court of Appeals supported the institution, but Buck's lawyers appealed, and the case came before the U.S. Supreme Court in 1927.

The Law and Biotechnology

Legal Issues

Buck's suit alleged that she had been deprived of the right of due process of law guaranteed under the Fourteenth Amendment. She also claimed to have been denied equal protection of the laws because the Virginia law affected people inside institutions but not those outside. The underlying issue was whether the state had a right to forcibly prevent reproduction by people it deemed to suffer from inherited defects and therefore to be likely to produce undesirable offspring. "It seems to be contended that in no circumstances could such an order be justified," Supreme Court Justice Oliver Wendell Holmes noted in his majority opinion.

Decision

The Supreme Court ruled that the Virginia eugenics law did not violate either the Due Process Clause or the Equal Protection Clause of the Fourteenth Amendment, and it therefore denied Buck's right to sue. Justice Holmes wrote that Buck had been granted due process because the Virginia law contained "very careful provisions . . . [to] protect the patients from possible abuse," including requirement of a hearing attended by both the inmate and his or her guardian to determine whether the person was "the probable potential parent of socially inadequate offspring." "There can be no doubt that . . . the rights of the patient are most carefully considered" in this procedure, Holmes claimed, and all the steps of it had been followed with "scrupulous" care in Buck's case.

Holmes also wrote that Buck had not been denied equal protection, even though the law did not affect all citizens of subnormal intelligence equally. "It is the usual last resort of constitutional arguments to point out shortcomings of this sort," he complained. However, he said, "the law does all that is needed when it does all that it can, indicates a policy, applies it to all within the lines, and seeks to bring within the lines all similarly situated so far and so fast as its means allow."

Perhaps most important, Holmes defended the social as well as the legal validity of the eugenics law. He wrote:

We have seen more than once that the public welfare may call upon the best citizens for their lives. It would be strange if it could not call upon those who already sap the strength of the State for these lesser sacrifices, often not felt to be such by those concerned, in order to prevent our being swamped with incompetence. It is better for all the world if, instead of waiting to execute degenerate offspring for crime or to let them starve for their imbecility, society can prevent those who are manifestly unfit from continuing their kind. The principle that sustains compulsory vaccination is broad enough to cover cutting the Fallopian tubes. Three generations of imbeciles are enough.

Impact

The court's decision in *Buck v. Bell* not only upheld the constitutionality of at least some eugenics laws but demonstrated the prevalent thinking that found such laws both scientifically and morally justified. The *Buck* case was cited often in subsequent decisions about similar laws, such as the case that follows.

SKINNER V. OKLAHOMA, 316 U.S. 535 (1942)

Background

Skinner, the plaintiff in this case, had not led an exemplary life. He was convicted of stealing chickens in 1926, followed by convictions for robbery with firearms in 1929 and 1934. In 1935, after his second armed robbery conviction, he was sentenced to the state penitentiary. Oklahoma law considered all three of Skinner's crimes to be "felonies involving moral turpitude" and stated that anyone convicted of two or more such felonies and sentenced to prison was a "habitual criminal." As such, the Habitual Criminal Sterilization Act, a 1935 Oklahoma eugenics law based on the belief that criminal tendencies were inherited, made him subject to sexual sterilization.

Skinner sued to prevent the operation. A jury trial confirmed that he could be sterilized without endangering his health. When the Oklahoma Supreme Court supported this decison, Skinner appealed on constitutional grounds. His case came before the U.S. Supreme Court in May 1942 and was decided on June 1.

Legal Issues

Skinner, like Carrie Buck before him, claimed that he had been denied equal protection under the Fourteenth Amendment. He also said that the Oklahoma law violated the Eighth Amendment because sterilization was cruel and unusual punishment. Underlying the particulars of the suit, as in the *Buck* case, was the question of whether eugenics laws as a whole were constitutional. The suit alleged that "the act cannot be sustained as an exercise of the police power, in view of the state of scientific authorities respecting inheritability of criminal traits."

The Law and Biotechnology

Decision

The Supreme Court ruled that Skinner had been denied equal protection because the Oklahoma law exempted embezzlers from the sterilization penalty but included those (like Skinner) who were convicted of grand larceny. The distinction between the two crimes was a very fine one, having to do with exactly when the convicted person had formed the intent of stealing. The state as a rule was entitled to make such fine distinctions, Justice William O. Douglas wrote in his majority opinion, but when a penalty as severe and permanent as sterilization was involved, they became much more dubious. Douglas wrote:

> *Strict scrutiny of the classification which a state makes in a sterilization law is essential, lest unwittingly, or otherwise, invidious discriminations are made against groups or types of individuals in violation of the constitutional guaranty of just and equal laws. . . . When the law lays an unequal hand on those who have committed intrinsically the same quality of offense and sterilizes one and not the other, it has made as invidious a discrimination as if it had selected a particular race or nationality for oppressive treatment.*

Douglas noted that there was no reason for assuming that a tendency to commit larceny was inheritable but a tendency to embezzle was not:

> *Oklahoma makes no attempt to say that he who commits larceny by trespass or trick or fraud has biologically inheritable traits which he who commits embezzlement lacks. . . . We have not the slightest basis for inferring that that line [between larceny and embezzlement] has any significance in eugenics.*

In contrast to the decision in the *Buck* case, Chief Justice Stone concluded in a concurring opinion that Skinner had been denied due process because the Oklahoma law, unlike the Virginia one, did not provide for a hearing in which an individual could present evidence that he or she is not "the probable potential parent of socially undesirable offspring." Stone accepted that "science has found . . . that there are certain types of mental deficiency associated with delinquency which are inheritable" and affirmed the right of the state to "constitutionally interfere with the personal liberty of the individual to prevent the transmission by inheritance of his socially injurious tendencies." He insisted, however, that there was no proof that the traits of any entire legal category of criminals were inheritable, and individuals therefore had the right to a hearing to determine whether their particular "criminal tendencies are of an inheritable type." Skinner, he said, had been denied that right. "A law which condemns, without hearing, all the individuals of a class to so harsh a measure [as sterilization] . . . because some or even many merit condemnation, is lacking in the first principles of due process."

The most important difference between the *Buck* and *Skinner* cases lay in the court's comments about the underlying social issue of eugenics and the government's right to forcibly prevent certain people from reproducing. Justice Douglas wrote, "This case touches a sensitive and important area of human rights . . . a right which is basic to the perpetuation of a race—the right to have offspring." The case, he said, "raised grave and substantial constitutional questions."

We are dealing here with legislation which involves one of the basic civil rights of man. Marriage and procreation are fundamental to the very existence and survival of the race. The power to sterilize, if exercised, may have subtle, far reaching and devastating effects. In evil or reckless hands it can cause races or types which are inimical to the dominant group to wither and disappear. There is no redemption for the individual whom the law touches. Any experiment which the state conducts is to his irreparable injury. He is forever deprived of a basic liberty.

Justice Jackson used equally strong words in a second concurring opinion.

I . . . think the present plan to sterilize the individual in pursuit of a eugenic plan to eliminate from the race characteristics that are only vaguely identified and which in our present state of knowledge are uncertain as to transmissibility presents . . . constitutional questions of gravity . . . There are limits to the extent to which a legislatively represented majority may conduct biological experiments at the expense of the dignity and personality and natural powers of a minority—even those who have been guilty of what the majority define as crimes.

Impact

Even though the court found for the plaintiff in this case, *Skinner* did not reverse the effects of *Buck*. It did not declare eugenics laws to be unconstitutional, scientifically invalid, or morally reprehensible. It did, however, express the sort of doubts about such laws that many people were beginning to feel. Its claim that reproduction was a basic right would often be cited in later cases.

Eugenics laws remained on the books of many states and countries until the 1970s. After the 1940s, however, they were seldom enforced. A combination of better understanding of heredity, which suggested that complex personality traits such as intelligence and criminality were determined as much by environment as by genetics, and a revulsion for eugenics principles triggered by revelation of Nazi genocide following World War II helped to make such laws unpopular.

NORMAN-BLOODSAW V. LAWRENCE BERKELEY LABORATORY, DOCKET 96-16526 (1998)

Background

While examining her medical records in the process of applying for workers' compensation in January 1995, an employee of Lawrence Berkeley Laboratory (LBL), a California research facility managed by the University of California and the U.S. Department of Energy, discovered that blood and urine samples she had given during a preemployment medical examination had been tested in several ways without her knowledge. The same proved to be true of other LBL employees.

After receiving a letter from the Equal Employment Opportunity Commission saying that they had grounds for a suit, Marya S. Norman-Bloodsaw and six other administrative and clerical employees of LBL filed suit against the laboratory and others in September 1995. The suit alleged that the laboratory had tested employees' blood and urine for syphilis, sickle-cell trait (in the case of black employees), and pregnancy (in the case of women). It was filed on behalf of all present and past Lawrence employees who had been subjected to the tests in question.

The U.S. District Court for the Northern District of California dismissed all the employees' claims in June 1997. They appealed, and the case went to the 9th Circuit Court of Appeals in February 1998.

Legal Issues

As Judge Stephen Reinhardt stated in the appeals court's written decision,

> *This appeal involves the question whether a clerical or administrative worker who undergoes a general employee health examination may, without his knowledge, be tested for highly private and sensitive medical and genetic information such as syphilis, sickle cell trait, and pregnancy.*

The LBL employees claimed that the medical tests in question had been administered without their knowledge or consent and that they were not notified later of the tests or their results. They said that their federal and state constitutional right to privacy had been violated by the conducting of the tests, the maintaining of the test results, and the lack of safeguards against disclosure of the results to others because of the "intimate" nature of the conditions tested. They claimed violations of Title VII of the Civil Rights Act of 1964 because only African American employees had been tested for sickle-cell trait and only women had been tested for pregnancy; furthermore, they alleged, later blood samples from black and Hispanic, but not other, employees had been tested again for syphilis. Finally, the employees claimed violations under the Americans with Disabilities Act because the tests were not related to their job performance or business necessity. They did not claim that LBL had taken any negative action regarding their jobs because of the tests or that it had revealed the test information to others, but they said that the laboratory had provided no safeguards against dissemination of that information.

In addition to asking for damages for themselves, the employees were suing, according to the court record,

> . . . to enjoin [forbid] future illegal testing, . . . to require defendants . . . to notify all employees who may have been tested illegally; to destroy the results of such illegal testing upon employee request; to describe any use to which the information was put, and any disclosures of the information that were made; and to submit Lawrence's medical department to "independent oversight and monitoring."

The defendants denied that any of the employees' claims had merit. The tests, they said, represented only a minimal intrusion beyond that which the employees had consented to as part of taking the medical examination and giving blood and urine samples. They claimed that signs posted in examination rooms, furthermore, had announced the tests and that employees had been asked about some of the items tested on a questionnaire that they completed as part of their examination. The questionnaire asked if the employees had ever had medical conditions including sickle-cell anemia, venereal disease, or (in the case of women) menstrual disorders. They therefore should not have been surprised at being tested for such conditions, the defendants claimed. The defendants also said that the testing had occurred so long ago that the statute of limitations for complaints about it had expired and that, in any case, the laboratory had stopped doing the syphilis tests in 1993 (because such tests turned out to be an economically inefficient way of screening a healthy population), pregnancy tests in 1994, and sickle-cell tests in 1995 (because most blacks were by then tested for sickle-cell trait at birth).

Decision

The circuit court of appeals reversed the district court's ruling that the statute of limitations prevented the plaintiffs from suing. The time limit began to run, the court said, from the time when the plaintiffs learned about the tests—1995—not the time when the tests were taken, as the district court had held. The circuit court said that the question of whether the plaintiffs knew or should have known that they were being tested would have to be settled at trial, but Judge Reinhardt maintained that the facts that the employees had consented to have a medical examination, give blood and urine samples, and answer written questions about certain medical conditions were "hardly sufficient" to establish an expectation of such testing. "There is a significant difference between answering [a questionnaire] on the basis of what you know about your health and consenting to let someone else investigate the most intimate aspects of your life," Reinhardt wrote. He also noted that "the record . . . contains considerable evidence that the manner in which the tests were performed was inconsistent with sound medical practice" in that the tests in question were not a routine or even an appropriate part of a standard occupational medical examination.

The appeals court upheld the district court's dismissal of the plaintiffs' claims under the Americans with Disabilities Act (ADA). First, Judge Reinhardt wrote, most of the testing at issue had occurred before January 26, 1992, the date on which the ADA began to apply to public entities. The employees tested after that date were tested as part of employee entrance examinations, which, unlike other examinations, are not required by the law to be limited to matters connected with a person's ability to perform job-related functions. The appeals court also disallowed claims under the ADA related to the way the employees' medical records were kept.

The appeals court supported the employees' right to sue on all the other grounds, however. It agreed that because of the tests' "highly sensitive" nature, they represented more than a minimal invasion of privacy beyond that involved in the medical examination that had been consented to. Judge Reinhardt wrote:

> *The constitutionally protected privacy interest in avoiding disclosure of personal matters clearly encompasses medical information and its confidentiality. . . . The most basic violation possible involves the performance of unauthorized tests—that is, the non-consensual retrieval of previously unrevealed medical information that may be unknown even to plaintiffs. These tests may also be viewed as searches in violation of Fourth Amendment rights. . . . The tests at issue . . . [also] implicate rights protected under . . . the Due Process Clause of the Fifth or Fourteenth Amendments. . . . One can think of few subject areas*

more personal and more likely to implicate privacy interests than that of one's health or genetic make-up.

The court ruled that discrimination in violation of Title VII of the Civil Rights Act of 1964 and the Pregnancy Discrimination Act was shown by the fact that certain tests were given to some employees but not to others as a condition of employment or were given more often to some employees during employment. The unauthorized obtaining of sensitive medical information on the basis of race or sex in itself constituted an "adverse effect" as defined by the act, even though no negative effects on employment occurred. The plaintiffs therefore had grounds to sue on this basis as well, the court decreed.

The fact that the tests had been discontinued did not make the plaintiffs' claims moot, the appeals court ruled, because the laboratory could reinstitute the tests at any time. Plaintiffs suffered ongoing injuries from the tests in that the test records were still in the employees' files and could potentially affect employment decisions or be given to others, even though this had not so far happened.

Impact

The decision confirmed employees' right to medical privacy and to not have tests, including tests for inherited conditions such as sickle-cell trait, run on them without their informed consent. In describing the decision, *U.S. News & World Report* writer Dana Hawkins said it represented "the first time a federal appeals court has recognized a constitutional right to genetic privacy."[18] The fact that Judge Reinhardt specifically mentioned genetic make-up in connection with privacy rights may prove particularly important to those concerned about privacy and discrimination related to genetic testing.

Partly because of this suit, Department of Energy contractors are now required to give employees a "clearly communicated" list of all medical examinations they will be expected to take, the purpose of the exams, and the results of the tests.

BRAGDON V. ABBOTT, 97 U.S. 156 (1998)

Background

In September 1994, Sidney Abbott visited her dentist, Randon Bragdon, in Bangor, Maine. In filling out a patient registration form, she indicated that she had been infected with HIV since 1986, although she had not yet developed any symptoms of AIDS. Bragdon examined Abbott, found that she had a cavity, and informed her that his policy was not to fill cavities of

HIV-positive patients in his office. He then offered to do the work at a nearby hospital (where he felt he could take better precautions to protect himself from infection) if Abbott was willing to pay the extra cost of using the hospital's facilities. She declined.

Abbott sued Bragdon for violating her rights under Title III of the Americans with Disabilities Act (ADA) by not treating her, since places of "public accommodation" defined in that section include the "professional office of a health care provider." Bragdon's lawyer pointed out that the act stated that people could refuse to do something for an individual that was otherwise required by the act if they could show that "said individual poses a direct threat to the health and safety of others," and he claimed that Abbott fit that description. Abbott's attorney disagreed, pointing out that the Centers for Disease Control and Prevention (CDC) and others had written guidelines describing procedures by which dentists could treat people with HIV infection.

A district court ruled that Abbott's HIV infection satisfied the requirements of disability under the ADA and that Bragdon had not proved that treating her would put his health at risk. The Court of Appeals affirmed both of the district court's rulings. Bragdon then appealed to the Supreme Court, and the case came before the court in June 1998.

Legal Issues

The court agreed to rule on the following points: (1) whether Abbott, as an asymptomatic person with HIV infection, was disabled as defined by the ADA, and (2) whether sufficient evidence had been provided to show that Bragdon's health would have been endangered by treating her.

The underlying issue of the case for those concerned about genetic discrimination was whether a person who was presently healthy but likely to become ill later could be considered disabled under the ADA, since this description fitted healthy people whom tests revealed to have a genetic susceptibility to a disease. A factor likely to affect this issue was the section, or "prong," of the ADA under which Abbott claimed disability. The act defines disability as having to meet one of three criteria:

1. a physical or mental impairment that substantially limits one or more of the major life activities of an individual;
2. a record of such impairment; or
3. being regarded as having such impairment.

Abbott claimed disability under the first prong, saying that HIV infection limited her in the major life activity of reproduction and childbearing. By contrast, the Equal Employment Opportunity Commission (EEOC) had stated as policy in 1995 that healthy people with genetic predispositions were covered under the act's third prong.

Decision

In a 5-4 decision, the Supreme Court ruled that Abbott's HIV infection rendered her disabled according to the ADA's first criterion, that of limitation of a major life activity. Justice Anthony Kennedy in his majority opinion stated that reproduction was, without question, a major life activity and that Abbott was substantially limited in her pursuit of it, since the unprotected sex necessary to conceive a child would put her partner at significant risk of infection. Furthermore, if she did become pregnant, the child would also have a good chance of being infected. Kennedy also noted that, even though HIV infection did not produce obvious symptoms for years, it caused steady damage to the blood and immune systems and thus was "an impairment from the moment of infection."

In writing of the possible threat to Bragdon's health, Justice Kennedy noted that the ADA defined a direct threat to be "a significant risk to the health or safety of others that cannot be eliminated by a modification of policies, practices, or procedures or by the provision of auxiliary aids or services." The basic question, Kennedy wrote, was whether Bragdon's actions were reasonable in light of the medical evidence available to him at the time. Kennedy pointed out that, on the one hand, Bragdon had not produced medical evidence to show that he would have been any safer treating Abbott in a hospital than in his office. On the other hand, the CDC and other similar guidelines do not necessarily say that dentists will be safe while treating HIV-infected patients if they follow the recommended procedures. Some such guidelines do say that risk is minimal if proper precautions are followed, and medical testimony had been offered to this effect as well in the previous trials, but Kennedy noted that Bragdon may not have had this information at the time he treated Abbott. Kennedy ordered the Court of Appeals to reconsider the evidence supporting Bragdon's estimation of his health risk.

Impact

According to the summary of a February 1999 workshop held jointly by the National Human Genome Research Institute and the Hereditary Susceptibility Group of the National Action Plan for Breast Cancer to discuss the implications of the *Bragdon v. Abbott* decision for healthy people diagnosed with a genetic susceptibility to breast cancer (and, presumably, other gene-related illnesses), the court's ruling "both excited and unnerved" those who hoped that the ADA could be interpreted in a way that would cover such people.[19] Chief Justice William Rehnquist, in fact, addressed this possibility when he wrote in his dissenting opinion that Abbott's argument, "taken to its logical extreme, would render every individual with a genetic marker for some debilitating disease 'disabled' here and now because of some possible future effects." On one hand, Rehnquist's words indicate that such reasoning is possible; on the other, he obviously was expressing disapproval of the idea.

In the workshop, law expert Paul Miller pointed out several hopeful signs in the *Bragdon* decision. The fact that the court considered HIV infection an actual disability from the beginning because of its effects on immune cells, even though no obvious symptoms of illness appeared, suggested that genetic tendencies, which may also cause physical or chemical changes in cells without producing visible symptoms, might be similarly classified. Furthermore, since reproduction had been affirmed as being a major life activity covered by the ADA, people with inherited defects could argue, as Abbott did, that their ability to reproduce was limited because they risked passing their condition on to their offspring and thus endangering those offspring's health. Third, Miller said, the court had relied heavily on an EEOC policy ruling in determining that asymptomatic HIV infection qualified as a disability under the ADA. This added weight to other EEOC rulings, including the one about genetic predisposition.

On the other hand, it was not clear how broad the effect of the court's ruling would prove to be. The ruling did not explicitly consider any life activity other than reproduction, for instance, so it might not cover, say, a postmenopausal woman or a gay man with asymptomatic HIV infection. More important, since Abbott claimed disability under the first prong of the ADA's definition, the court decision did not illuminate the question of who would be included under the third prong, which many commentators feel is the one most likely to cover healthy people with genetic predispositions. Miller noted that many legislators are not supportive of the third prong, and a later speaker, Sharon Masling, said that the courts also have been interpreting it narrowly.

Further doubt about the possibility of including healthy people with genetic predispositions in the ADA's protected classes has been raised by a June 1999 Supreme Court decision on a group of four other ADA cases. The court ruled that the ADA does not cover people whose disabilities can be corrected, such as people with poor eyesight who wear glasses. In her majority opinion on these cases, Justice Sandra Day O'Connor wrote regarding the ADA requirement that a disability substantially limit a major life activity that "We think the language [of the ADA] is properly read as requiring that a person be presently—not potentially or hypothetically—substantially limited in order to demonstrate a disability." O'Connor's line of reasoning would seem to make ADA coverage unlikely for people with genetic predispositions who are not presently ill.

1 Mitchel Zoler, quoted in Martin Kenney, *Biotechnology: The University-Industrial Complex* (New Haven: Yale University Press, 1986), p. 256.

2 Donald Dunner, quoted in Kenney, p. 256.

3 Peter Farley, quoted in Edward J. Sylvester and Lynn C. Klotz, *The Gene Age: Genetic Engineering and the Next Industrial Revolution* (New York: Scribner, 1983), p. 118.

4 Office of Technology Assessment (1987), quoted in *John Moore v. Regents of California*.

5 Unknown juror, quoted in Peter J. Neufeld and Neville Coleman, "When Science Takes the Witness Stand," *Scientific American*, May 1990, p. 46.

6 Peter Neufeld, quoted in Roger Lewin, "DNA Typing on the Witness Stand," *Science*, June 2, 1989, p. 1033.

7 Lewin, p. 1033.

8 Richard Roberts, quoted in Lewin, p. 1033.

9 Eric Lander, "DNA Fingerprinting: Science, Law, and the Ultimate Identifier," in Daniel J. Kevles and Leroy Hood, eds., *The Code of Codes: Scientific and Social Issues in the Human Genome Project* (Cambridge, Mass.: Harvard University Press, 1992), p. 201.

10 Richard Roberts, Eric Lander, et al., quoted in Lander, p. 201.

11 Peter Neufeld, quoted in Lewin, p. 1033.

12 Eric Lander, quoted in Shawna Vogel, "The Case of the Unraveling DNA," *Discover*, January 1990, p. 46.

13 Jamie Gorelick, quoted in "Fugitive Justice," *Nation*, March 3, 1997, p. 4.

14 Edward Imwinkelried, quoted in Lewin, p. 1035.

15 Richard Roberts, quoted in Lewin, p. 1035.

16 James Carr, quoted in Sylvester and Klotz, p. 3.

17 Daniel Hartl, quoted in Rachel Nowak, "Forensic DNA Goes to Court with O. J.," *Science*, September 2, 1994, p. 1353.

18 Dana Hawkins, "Court Declares Right to Genetic Privacy," *U.S. News & World Report*, February 16, 1998, p. 4.

19 National Action Plan for Breast Cancer, "*Bragdon v. Abbott*: Indications for Asymptomatic Conditions." Available online. URL: http://www.napbc.org/napbc/heredita1.htm. Posted February 1999.

CHAPTER 3

CHRONOLOGY

This chapter presents a chronology of important events relevant to biotechnology, primarily the subset of biotechnology that involves genetic engineering. It also includes key events in the history of genetics, since genetic engineering would have been impossible without the knowledge gained from basic genetic research. It focuses primarily on the period following the invention of genetic engineering in the early 1970s and on events related to the ethical, legal, and social implications of biotechnology, genetic engineering, and human genetics.

ABOUT 10,000 YEARS AGO

- Biotechnology begins along with agriculture. It includes domestication and deliberate breeding of plants and animals, as well as (unknowing) use of microbial processes to make bread, cheese, alcoholic drinks, leather, and other products.

1665

- British scientist Robert Hooke discovers microscopic square bodies in a slice of cork and terms them cells.

1839

- German biologists Matthias Schleiden and Theodore Schwann propose that cells are the units of which all living things are made.

LATE 1850s–1860s

- Famed French chemist Louis Pasteur provides a scientific basis for part of traditional biotechnology when he shows that fermentation processes such as those used for millennia to produce wine, beer, buttermilk, and cheese depend on living microorganisms.

1859

■ British biologist Charles Darwin publishes *On the Origin of Species*, in which he sets forth the theory of evolution by natural selection. The theory states that the members of a species with inherited characteristics that make them most suited to survive in a particular environment are the most likely to survive and bear young. Over generations, therefore, a change in the environment can cause a change in the predominant characteristics of a species.

1866

■ Gregor Mendel, an Austrian monk, publishes a paper describing the basic mathematical rules by which characteristics are inherited. He had worked these out by breeding pea plants in his monastery garden.

1875

■ German scientist Walther Flemming discovers that the nuclei (central bodies) of cells contain stringlike bodies that can be stained with dye; these are soon termed chromosomes, or "colored bodies."

1883

■ British scientist Francis Galton coins the term *eugenics* (from the Greek meaning "well born") in a book called *Inquiry into Human Faculty*; he and his followers believe that the human race can be improved by encouraging those with desirable traits to have children and discouraging or preventing those with undesirable traits from doing so.

1900

■ Several scientists rediscover Mendel's work, which until this time has been virtually unknown; it now becomes the foundation of genetics.

1910

■ American geneticist Thomas Hunt Morgan and coworkers at Columbia University prove that genes are located on chromosomes.

Chronology

1920s–1930s

- Thirty-four states in the United States pass eugenics laws requiring people with what are thought to be inherited defects (chiefly criminals, the insane, and the "feebleminded") to be forcibly sterilized.

1927

- In *Buck v. Bell*, the U.S. Supreme Court by an 8-1 vote upholds a Virginia eugenics law under which Carrie Buck, a developmentally disabled woman, was forcibly sterilized. Noting that Buck's mother and seven-month-old daughter both appear to be "feebleminded" as well, Justice Oliver Wendell Holmes writes, "Three generations of imbeciles are enough."

1930

- U.S. Plant Patent Act allows breeders to protect plant varieties they have developed by refusing to allow others to reproduce such plants asexually.

1933

- The German government, recently seized by the Nazi party, passes eugenics law modeled on those of the United States.

1938

- U.S. Congress passes the Federal Food, Drug, and Cosmetic Act (FFDCA), which gives the Food and Drug Administration (FDA) the right to set tolerances for certain substances, including pesticides, on or in food and feed. The FDA now shares this regulatory duty with the Environmental Protection Agency (EPA). Herbicide residues on genetically engineered herbicide-tolerant crops and pesticides (such as Bt toxin) produced in genetically altered food crops are regulated under the FFDCA.

1941

- George Beadle and Edward L. Tatum of Stanford University show that a single (structural) gene carries the instructions for making a single protein (enzyme).

1942

- In *Skinner v. Oklahoma*, the U.S. Supreme Court strikes down a state law requiring forced sterilization of convicted criminals, saying that procreation is "one of the basic civil rights of man."

Biotechnology and Genetic Engineering

1944

- Oswald Avery and colleagues at the Rockefeller Institute in New York publish a paper demonstrating that DNA, not protein, carries inherited information.

1952

- U.S. Congress revises patent law to state that patentable inventions must be novel, useful, and not obvious "to a person of ordinary skill in the art."

1953

- *April 25*: James D. Watson and Francis Crick publish a paper in *Nature* in which they describe the structure of the DNA molecule.

- *May*: Watson and Crick publish a second paper offering a theory that explains how DNA's structure allows the molecule to reproduce itself.

1958

- Watson and Crick's theory of how the DNA molecule reproduces itself is confirmed.

1961

- Francis Crick suggests that each "letter" of the genetic code—the unit specifying one amino acid in a protein—consists of three adjoining bases in a DNA molecule.

1961–1966

- Molecular biologists decipher the genetic code, determining which amino acid each of the 64 possible combinations of three bases represents. They also work out the process by which the cell makes the proteins specified by the code.

1970

- U.S. Plant Variety Protection Act allows breeders to protect new plant varieties by forbidding others to reproduce the plants sexually.

1971

- Robert Pollack warns Paul Berg about the possible dangers of inserting genes from a cancer-causing virus into bacteria capable of infecting humans.

Chronology

1972

- U.S. Congress passes the Sickle-Cell Anemia Act, which sets up a screening program to identify sickle cell carriers but specifies that participation in the program must be voluntary and must not be linked to eligibility for federal services.

- *November*: Herbert Boyer and Stanley Cohen meet in a Hawaiian delicatessen and begin planning recombinant DNA technology.

1972–1973

- Paul Berg, Herbert Boyer, and Stanley Cohen perform the first experiments in which pieces of DNA from one species are inserted into the genome of another species (recombinant DNA).

1973–1974

- Leading scientists in the field write two letters, published in *Science*, that warn of possible dangers of recombinant DNA experiments and call for a moratorium on some types of experiments.

1974

- Herbert Boyer and Stanley Cohen apply for a patent on their gene-splicing technique and assign all potential royalties from it to their respective universities (University of California at San Francisco and Stanford University).

- *October*: The National Institutes of Health establishes the Recombinant DNA Advisory Committee (RAC) to review the safety of recombinant DNA experiments.

1975

- North Carolina passes a law forbidding employers to discriminate against people with sickle-cell trait. This is the first American law to address genetic discrimination in the workplace.

- *February 24–27*: One hundred forty geneticists and molecular biologists meet at Asilomar, California, to draw up safety guidelines for recombinant DNA experiments.

- *April*: Congress holds first hearings about safety of recombinant DNA research.

Biotechnology and Genetic Engineering

1976

- Robert Swanson and Herbert Boyer found Genentech, the first biotechnology company based on recombinant DNA technology.

- U.S. Congress passes the Toxic Substances Control Act (TSCA), which gives the Environmental Protection Agency (EPA) the authority to review new chemicals before they are introduced into commerce. TSCA is later amended to classify some genetically engineered organisms as "new chemicals" to be regulated under the act.

- Congress passes the Genetic Diseases Act, which provides research, professional education, screening, and counseling related to inherited diseases; it requires screening for carriers of such diseases to be voluntary.

- *June*: National Institutes of Health (NIH) publishes safety guidelines for recombinant DNA experiments, based on the Asilomar guidelines. Britain and Europe adopt similar guidelines.

- *July*: The city council of Cambridge, Massachusetts, imposes a moratorium on recombinant DNA experiments within the city's borders.

1977

- Sixteen bills regulating recombinant DNA research are introduced into Congress; none of them passes.

1978

- NIH guidelines are relaxed, reducing containment requirements for many recombinant DNA experiments.

- Genentech makes human insulin in genetically engineered bacteria.

- David Botstein and others develop restriction fragment length polymorphism (RFLP) analysis, which will later be used to locate genes that cause inherited diseases and to identify DNA from particular individuals.

- J. B. Lippincott publishes *In His Image: The Cloning of a Man*, by science writer David Rorvik, in which Rorvik claims to have witnessed the cloning of an anonymous wealthy man. The book is later shown to be a hoax.

1980

- NIH safety guidelines for recombinant DNA research are relaxed further.

- Martin Cline of the University of California, Los Angeles, attempts human gene therapy overseas after being denied permission for such therapy by the RAC. The treatment fails to get altered genes into Cline's patients.

- *June 16*: In *Diamond v. Chakrabarty*, a landmark case, the U.S. Supreme Court declares that living things can be patented if humans have altered them.

- *October*: Genentech stock is offered to the public for the first time and jumps from $35 to $89 a share in the first few minutes of trading despite the fact that the company has not yet sold any products.

EARLY 1980S

- First transgenic plants produced.

1982

- The FDA grants Genentech permission to sell genetically engineered human insulin.

- Steven Lindow applies to the RAC for permission to test genetically altered *Pseudomonas* bacteria in open fields.

1983

- Kary Mullis discovers the polymerase chain reaction (PCR), which can be used to duplicate small amounts of DNA many times in a few hours.

1985

- The first transgenic farm animals are created.

- Alec Jeffreys of the University of Leicester, Great Britain, publishes a paper in *Nature* describing what he calls "DNA 'fingerprinting,'" a way to use genetic testing to identify individuals.

- *October*: The first United States patent for a genetically altered plant is issued.

1986

- First mammals (sheep) cloned by nuclear transfer technique, using embryonic cells.

- The Toxic Substances Control Act (TSCA) is amended to require an EPA permit for releasing genetically altered organisms into the environment.

- *June*: U.S. Office of Science and Technology Policy issues framework for regulation of biotechnology, dividing such regulation among five government agencies.

1987

- U.S. Congress passes the Plant Pest Act, which gives the U.S. Department of Agriculture's Animal and Plant Health Inspection Service (APHIS) the right to regulate any organisms, including those produced by genetic engineering, that are or might be plant pests.

- First people (Colin Pitchfork in Britain and Tommie Lee Andrews in the United States) convicted of crimes on the basis of DNA identification testing.

- W. French Anderson asks the RAC for permission to try gene therapy against ADA deficiency and is refused.

- *April*: The U.S. Patent and Trademark Office extends patentability to animals, cell lines, and genes, including human cells and genes.

- *April 24*: Steven Lindow and coworkers oversee spraying of "ice-minus" bacteria onto strawberry plants in a field in Contra Costa County, California. This is the first deliberate release of genetically engineered organisms into the environment.

1988

- *April*: The Harvard Oncomouse, a mouse genetically altered to make it unusually susceptible to cancer and thus useful in cancer research and carcinogen testing, becomes the first genetically engineered animal to be patented.

1989

- During a pretrial hearing for *New York v. Castro*, geneticists called as expert witnesses for both sides meet outside the courtroom and agree that DNA tests intended to provide evidence in the case have been so poorly conducted that they make the evidence worthless.

- *May 22*: Steven Rosenberg and coworkers at the National Institutes of Health carry out the first successful insertion of foreign genes into a human being. The genes are intended only as markers and have no effect on health.

Chronology

1990

- The Human Genome Project, which aims to sequence all 100,000 or so human genes by a few years into the 21st century, begins.

- U.S. Congress passes the Americans with Disabilities Act, which may ban discrimination against healthy people with inherited defects revealed by genetic tests.

- *Summer*: In a pretrial hearing in *United States v. Yee*, geneticists and lawyers argue about the accuracy of the method used to calculate the statistics of two people's DNA test profiles matching by chance.

- *July*: The California Supreme Court rules in *John Moore v. Regents of University of California* that people do not retain ownership of their cells or tissues once these have been removed from their bodies.

- *September 14*: French Anderson and colleagues from the NIH administer the first successful gene therapy to four-year-old Ashanthi deSilva.

1991

- Working at the NIH, Craig Venter develops a method to use so-called expressed sequence tags (ESTs) as probes to find unknown genes. NIH tries to patent several thousand ESTs, drawing protest.

- The FBI establishes guidelines for forensic DNA testing.

- James Watson, head of the Human Genome Project, earmarks 3 percent of the project's budget for investigation of its ethical, legal, and social implications.

- The National Biodiversity Institute (INBio) of Costa Rica promises to provide samples of the country's wild plants, microbes, and insects to drug giant Merck in return for a large research and sampling budget, training, and royalties from any resulting products; some of the money will be used to preserve the country's national parks.

- *December 20*: Landmark articles appear in *Science* presenting opposing viewpoints about the validity of the assumptions underlying the method used to calculate the probability of DNA identification profiles from two people matching by chance.

1992

- The PTO grants a patent on all genetically engineered cotton plants, regardless of how they are produced, to Agracetus.

- Human Genome Diversity Project established.

- U.S. Department of Defense begins requiring all military personnel to give DNA samples, which will be retained in a database to help identify remains of soldiers killed in action.

- *May*: The FDA states that genetically altered foods do not have to receive special approval or be labeled as such if they are nutritionally the same as their natural equivalents and contain no new substances that might cause an allergic reaction.

- The National Research Council of the National Academy of Sciences issues a report on DNA identification that attempts (but fails) to end the controversy about the validity of DNA profiling statistics.

1993

- The FDA approves recombinant bovine growth hormone (rBGH) for sale; it is the first genetically engineered animal hormone approved for sale in the United States.

- Protests force NIH to drop its application to patent a cell line derived from a Guaymi (Panama) Indian woman.

- *March*: The USDA streamlines its requirements for testing new varieties of genetically engineered corn, cotton, soybeans, potatoes, tomatoes, and tobacco to eliminate a former requirement for an environmental review before testing.

- *October*: Robert Stillman and Jerry Hall of George Washington University Medical Center in Washington, D.C., clone early-stage human embryos. The embryos, already due to be discarded by a fertility clinic, are not allowed to develop.

1994

- The Flavr Savr tomato becomes the first genetically engineered food to go on sale.

- The NIH applies for a patent on a cell line from a man of the Hagahai people in Papua New Guinea. The patent is granted in 1995, but protests cause NIH to give it up in 1996.

- Congress passes the DNA Identification Act, which gives the FBI national responsibility for training, funding, and proficiency testing of laboratories doing forensic DNA profiling.

- China passes a law requiring marriage bans, forced sterilizations, or abortions to prevent births of children with inherited diseases.

- The Equal Employment Opportunity Commission (EEOC) rules that denying employment to people who are healthy but have inherited defects revealed by genetic tests violates the Americans with Disabilities Act.

1995

- *May*: A coalition of more than 200 religious leaders, organized by Jeremy Rifkin, presents a petition opposing the patenting of animals or humans or their organs, tissues, or genes.

- *October*: Former football star O. J. Simpson is acquitted of the murder of his wife and an acquaintance, Ronald Goldman. The verdict comes despite DNA evidence linking Simpson to the crimes, after his defense lawyers show that samples from the crime scene were mishandled and may have been accidentally or deliberately contaminated with Simpson's blood before testing.

1996

- Genetically altered crops, including those that resist herbicides and those that produce their own insecticide, are sold commercially and planted on a large scale in the United States for the first time.

- Affymetrix invents the GeneChip (also called DNA chip or biochip), a computer chip coated with DNA that allows thousands of genes to be scanned for mutations at a time.

- Several studies reveal evidence of discrimination in insurance and employment on the basis of the results of genetic tests.

- The NRC issues a second report on forensic DNA testing that withdraws its controversial compromise method of calculating match statistics and its call for independent proficiency testing of forensic DNA laboratories.

- U.S. Congress passes the Health Insurance Portability and Accountability Act, which includes a provision forbidding health insurers issuing group plans to deny coverage to healthy people because of preexisting genetic conditions.

- *April*: U.S. Marines John C. Mayfield III and Joseph Vlakovsky are court-martialed and found guilty of disobeying a direct order because they refused to give DNA samples for storage in a military database, saying that doing so invaded their genetic privacy.

1997

- Multinational giant Monsanto Corporation spins off its chemical division and reorganizes itself as a "life sciences company" emphasizing biotechnology, especially in foods, agricultural products, and pharmaceuticals.

- *January*: Deputy Attorney General Jamie Gorelick publicly admits that the FBI crime laboratory has a "serious set of problems" in its handling and testing of forensic DNA samples.

- *February 27*: Ian Wilmut and coworkers at the Roslin Institute in Scotland publish an article in *Nature* announcing that they have cloned a sheep (Dolly) from a mature udder cell taken from a six-year-old (adult) ewe.

- *March 4*: Reacting to fears stirred by the news about Dolly, President Bill Clinton announces a ban on the use of federal funds for research on cloning of human beings.

- *March 5*: The first Congressional hearing on human cloning takes place before the House Science Committee's Technology Subcommittee.

- *March*: Senator Christopher Bond and Representative Vernon Ehlers introduce bills that will ban human cloning.

- *June*: A report from the National Bioethics Advisory Commission recommends a legislative moratorium on human cloning for three to five years.

- *July 16*: European Parliament approves draft directive on intellectual property rights in biotechnology.

- *November*: The European Union begins requiring labeling of all genetically modified foods.

- *November*: The Council of Europe adopts a convention concerning human rights and dignity in regard to biology and medicine, which includes a provision forbidding discrimination on grounds of genetic makeup.

1998

- Two private genetics companies claim that they will finish sequencing the human genome before the Human Genome Project does.

- *January*: President Clinton calls for laws to protect against genetic discrimination by insurers; Vice President Gore makes a similar request regarding genetic discrimination in the workplace.

Chronology

- *January*: Chicago physicist Richard Seed revives concerns about human cloning when he announces that he will clone human children in the next few years.

- *January*: California becomes the first state to outlaw human cloning.

- *January 12*: The Council of Europe signs a ban on human cloning.

- *February*: In *Norman-Bloodsaw v. Lawrence Berkeley Laboratories*, the U.S. Court of Appeals for the 9th Circuit rules that Lawrence Berkeley Laboratory's genetic and other testing of employees' blood and urine samples without their informed consent was unconstitutional.

- *February*: Lobbying by scientists and biotechnology industry representatives blocks the Republican anti-human cloning bill, which had been sent to the Senate floor. Democrats introduce an alternative bill that, unlike the Republican one, permits research on cloned human embryos that are not allowed to develop.

- *June*: In *Bragdon v. Abbott*, the Supreme Court rules that the Americans with Disabilities Act (ADA) protects to a healthy woman who is HIV positive, leading to speculation that it may also apply to healthy people with genetic problems revealed by testing.

- *October*: The FBI opens its Combined DNA Index System (CODIS), a national database that will store information from DNA samples taken from convicted murderers and sex offenders by the states.

- *November*: Geron Corp. announces that it has isolated and cultivated human embryonic stem cells, which potentially can provide tissues for transplantation. Further research on these cells may involve the use of cloned human embryos.

1999

- *March*: Attorney General Janet Reno asks a federal commission to consider the constitutionality of requiring DNA samples from anyone arrested for a crime, whether or not the person is later convicted.

- *May*: A Cornell University study suggests that pollen from corn crops genetically altered to produce an insecticide may land on milkweed plants and poison monarch butterflies.

- *May 23*: Roslin Institute scientists announce that, at least according to one measurement, the cells of Dolly the cloned sheep match the age of her donor "mother" rather than her own birth age.

- *May 23*: The National Bioethics Advisory Commission calls for lifting the ban on federal funding for research involving human embryos because of the medical importance of research on embryonic stem cells.

- *June*: In a Supreme Court decision on several ADA cases, Justice Sandra Day O'Connor writes that the law protects only those who are presently disabled, dimming hopes that the ADA could be applied to healthy people with genetic predispositions to disease.

- *July 1*: The House of Representatives passes a bill that would allow health insurers to reveal genetic and other medical information to credit card companies.

- *July 13*: Secretary of Agriculture Dan Glickman announces that the United States will conduct long-term studies on the safety to health and environment of genetically modified farm crops for the first time. It will also review its procedure for approving such crops and encourage the industry to label genetically engineered products. At the same time, Glickman says, the United States will use all available legal means, including tariffs on European-made goods, to compel Europe to accept American farm products, including those that may contain bioengineered materials.

- *September 17*: An eighteen-year-old Arizona man with a rare inherited disease dies after taking part in a gene therapy experiment. His is the first death officially attributed to gene therapy.

- *November 2*: Researchers and drug companies admit having concealed from the NIH six deaths during a different gene therapy experiment. They claim the therapy did not cause the deaths.

- *December*: Protesters raise issue of genetically modified foods during World Trade Organization talks in Seattle.

- *December 13*: Food and Drug Administration holds hearing in Oakland, California, about whether to institute stricter safety testing and mandatory labeling for genetically modified food.

2000

- *June 26*: Teams of scientists from the private company Celera Genomics and the government-funded National Human Genome Research Institute announce that they have finished the first rough draft of the code of the complete human genome.

CHAPTER 4

BIOGRAPHICAL LISTING

This chapter offers brief biographical information on people who have played major roles in the development of biotechnology and genetic engineering or in events relating to the ethical, legal, and social implications of these fields. Some pioneers of genetics and molecular biology are also included, but the list focuses primarily on people who have been active since the development of genetic engineering in the early 1970s.

Sidney Abbott, an HIV-positive woman with no symptoms of AIDS. She sued her dentist, Randon Bragdon, for refusing to treat her, saying that his refusal violated her rights under the Americans with Disabilities Act (ADA). In June 1998 the Supreme Court confirmed her right to sue, leading to hope that the ADA might also cover healthy people with a genetic predisposition to disease.

W. French Anderson, pioneer in gene therapy. He headed the National Institutes of Health (NIH) team that made the first successful use of altered genes to treat a human disease in September 1990. He is currently at the University of Southern California. He plans to use gene therapy on fetuses with gene defects within a few years.

Lori B. Andrews, professor at Chicago-Kent College of Law and director of the Institute of Science, Law, and Technology at the Illinois Institute of Technology. She is an expert on reproductive technologies and the law and a former head of the Human Genome Project's Ethical, Legal, and Social Issues Working Group. She has written a book on possible human cloning and other new reproductive technologies called *The Clone Age: Adventures in the New World of Reproductive Technology* (1999).

Tommie Lee Andrews, the first American convicted of a criminal charge primarily through DNA identification testing. Andrews, a 24-year-old warehouse worker, was convicted of the rape and beating of several women in Orlando, Florida, in late 1987 and early 1988 after the DNA profile from a sample of his blood matched that of semen found on one of the victims.

Oswald Avery, Canadian-born bacteriologist who proved that DNA rather than proteins carries inherited information. In 1944, Avery and his colleagues at the Rockefeller Institute in New York published a paper showing that a form of *Pneumococcus* bacteria incapable of causing disease became able to cause illness after they took up DNA from a disease-causing form of the same bacteria. The change was passed on to the bacteria's descendants.

George Beadle, Stanford University molecular biologist, showed in 1941 that each gene in a DNA molecule coded for one enzyme (that is, a protein molecule). He and his colleague, Edward L. Tatum, won a Nobel Prize for this work in 1958.

Paul Berg, Stanford University biochemist sometimes called "the father of genetic engineering." Berg did the first recombinant DNA experiments, transferring genes between two types of viruses, in the winter of 1972–73. Berg was also one of the first scientists to question the safety of recombinant DNA experiments.

Paul R. Billings, expert on discrimination resulting from genetic health testing. Billings, who is affiliated with the Veterans Administration Health Administration, has published several studies showing examples of healthy people who were denied health insurance or employment because testing revealed that they or members of their families carried disease-related genes.

R. Michael Blaese, leader, with W. French Anderson, of the first successful application of gene therapy to human disease in 1990. Blaese, like Anderson, worked for the National Institutes of Health; he was an expert on adenosine deaminase (ADA) deficiency, the rare inherited condition to which the therapy was applied. He now works for Kimeragen, a biotechnology company.

Christopher (Kit) Bond, Republican senator from Missouri. Bond cosponsored a bill to ban human cloning that was introduced to the Senate in 1997 but failed to pass. It was reintroduced in 1998 with the support of Senate Majority Leader Trent Lott and Bill Frist, the Senate's only physician member, after physicist Richard Seed claimed that he would soon begin cloning humans. It had not passed by 1999.

Herbert Boyer, a biochemist. In 1973, while at the University of California at San Francisco, he worked with Stanley Cohen of Stanford to develop the first practical technique for transferring genes between organisms. Boyer and venture capitalist Robert Swanson cofounded Genentech, the first business built on recombinant DNA technology, in 1976.

Carrie Buck, a developmentally disabled woman in an institution who appealed a decision to forcibly sterilize her under a Virginia eugenics law.

The U.S. Supreme Court upheld the law by an 8-1 vote in 1927. Noting that Buck's mother and seven-month-old daughter as well as Buck herself appeared to be "feebleminded," Justice Oliver Wendell Holmes wrote in the majority opinion on the case that "three generations of imbeciles are enough."

Thomas Caskey, director of the Institute for Medical Genetics at the Baylor College of Medicine in Houston, Texas. Caskey testified for the prosecution in the contentious pretrial hearing in *United States v. Yee* in the summer of 1990. In the mid 1990s he invented the Short Tandem Repeat (STR) method of DNA testing, a currently preferred method that combines elements of both older techniques (RFLP and PCR).

Luca Cavalli-Sforza, Stanford geneticist and head of the Human Genome Diversity Project. This project aims to collect DNA samples from about 500 of the world's most endangered indigenous peoples and study them to learn about such subjects as past human migrations and differences in susceptibility to disease. Some indigenous people's groups have claimed that the project is insensitive to native rights and beliefs and will result in exploitation of their genetic heritage.

Ananda Chakrabarty, General Electric Corporation scientist. He obtained the first United States patent on a living thing. His application for a patent on a genetically engineered bacterium that digested petroleum was rejected at first by the Patent and Trademark Office, but in 1980 the Supreme Court ruled by a 5 to 4 vote that Congress had intended "anything under the sun that is made by man" to be patentable, including altered organisms such as the one Chakrabarty had invented.

Ranajit Chakraborty, geneticist at the University of Texas, Houston. Chakraborty was coauthor, with Kenneth Kidd, of a December 1991 article in *Science* rebutting the claim of population geneticists Richard Lewontin and Daniel Hartl that the method used to calculate the probability of a chance match between samples from two people in DNA tests was inaccurate.

Martin Cline, scientist at the University of California, Los Angeles. In 1980 he made the first attempt at human gene therapy. Denied permission by the Recombinant DNA Advisory Committee (RAC) to try his treatment for an inherited blood disease called thalassemia in the United States, Cline carried out his procedure in Italy and Israel. It failed, causing many scientists to conclude that human gene therapy was not practical.

Stanley Cohen, Stanford University geneticist. With Herbert Boyer, he invented the first practical method of transferring genes between organisms in 1973. Cohen had studied plasmids, small rings of DNA that bacteria sometimes use to transfer genetic information, and he and Boyer worked out a method of combining plasmids from two types of bacteria.

Biotechnology and Genetic Engineering

Francis Collins, a molecular biologist. He currently directs the National Institutes of Health's National Human Genome Research Institute and the Human Genome Project.

Francis Crick, British molecular biologist. With James Watson, he worked out the structure of the DNA molecule at Cambridge University and showed how this structure allowed the molecule to reproduce itself. Watson, Crick, and Maurice Wilkins shared a Nobel Prize for this work in 1962. In the early 1960s, Crick set out a theory explaining how inherited information was coded in DNA and how the cell translated this information into proteins.

Kenneth Culver, a member of the National Institutes of Health team that carried out the first successful human gene therapy in September 1990. In more recent years, Culver has experimented with inserting a virus gene into human brain tumors to make them susceptible to an antiviral drug. He has been executive director of the Human Gene Therapy Research Institute at Iowa Methodist Medical Center.

Charles Darwin, author of the theory of evolution by natural selection, which he set forth in *On the Origin of Species* (1859). The theory states that the members of a species with inherited characteristics that make them most suited to survive in a particular environment will be the most likely to survive and bear young. Over generations, therefore, a change in the environment can cause a change in the predominant characteristics of a species.

Ashanthi (Ashi) deSilva was four years old in September 1990, when she made history as the first person to receive altered genes as a treatment for disease. Ashanthi had inherited a rare condition called ADA (adenosine deaminase) deficiency, which left her without a functioning immune system. After several treatments in which her own blood cells were reinjected after being given normal ADA genes in the laboratory, she became healthy enough to lead an essentially normal life.

Vernon J. Ehlers, a Republican member of the House of Representatives from Michigan. An outspoken foe of human cloning, he submitted two anticloning bills to the House in 1997, one of which passed in the House Science Committee.

Dianne Feinstein, Democratic senator from California. She cosponsored (with Ted Kennedy) a Senate bill to ban human cloning that was introduced in early 1998. Biotechnology industry spokespeople and many scientists preferred this bill to a Republican one with the same purpose because, unlike the Republican bill, the Feinstein-Kennedy bill did not ban research on cloned human embryos.

Walther Flemming, a German biologist. In 1875, he discovered chromosomes ("colored bodies") in the nuclei or central bodies of cells and showed how these changed as cells reproduced.

Biographical Listing

Francis Galton, a cousin of Charles Darwin. He founded the pseudoscience of eugenics, which held that personality characteristics such as intelligence or laziness are inherited and that the human race can be improved by encouraging people with desirable characteristics to have children and discouraging reproduction in those with undesirable characteristics. Some of Galton's supporters advocated sterilization by force. Eugenics has fallen into disrepute, but Galton should also be remembered as the founder of biostatistics (application of statistics to animal populations) and of identification by fingerprinting.

David Golde, physician at the University of California, Los Angeles, (UCLA) Medical Center, was a defendant in a suit concerning ownership rights to tissue that was made into a commercial cell line. John Moore, whom Golde had treated for leukemia, sued Golde and others in 1984 because they had developed a lucrative cell line from Moore's spleen (removed as part of his cancer treatment) without obtaining Moore's permission or offering him any recompense. The California Supreme Court ruled in 1990 that Moore had no ownership rights to his tissue but that Golde had violated his "fiduciary duty" to Moore by not telling him about his commercial plans.

Jerry Hall, a scientist at George Washington University Medical Center in Washington, D.C. With Robert Stillman, also at George Washington, he made headlines in October 1993 by successfully cloning early-stage human embryos. Stillman and Hall used embryos that had been scheduled for destruction at a fertility center and did not allow them to develop into fetuses, but opponents of human cloning and of embryo research nonetheless protested their work.

Daniel Hartl, a population geneticist. He coauthored a 1991 article with Richard Lewontin in which they maintained that the usual method of calculating the probability that DNA samples from two people would match by chance was inaccurate because it did not take into account the fact that large population groups such as Caucasians were made up of ethnic subpopulations and that people usually married within their subpopulation. Hartl also testified for the defense in the *United States v. Yee* pretrial hearing in the summer of 1990.

Oliver Wendell Holmes, Supreme Court Justice from 1902 to 1932. In the 1927 case *Buck v. Bell*, Holmes wrote the majority opinion that supported a Virginia eugenics law under which a developmentally disabled woman, Carrie Buck, was forcibly sterilized. Noting that not only Buck but her mother and her seven-month-old daughter were of subnormal intelligence, Holmes wrote that "three generations of imbeciles are enough."

Leroy Hood, a pioneer inventor of machines that handle DNA automatically. In 1983 he and coworkers at the California Institute of Technology invented a machine that could assemble short stretches of DNA of known sequence. They invented one that could automatically determine the sequence of stretches of DNA in 1986. Hood's machines have helped make gargantuan tasks like the Human Genome Project possible.

Robert Hooke, British scientist, inventor, and philosopher. In addition to many achievements in chemistry and physics, Hooke invented or greatly improved the compound microscope. In 1665, using this invention, he observed tiny, square structures in a slice of cork and dubbed them cells. In fact, what he saw were merely the walls that were all that remained of these basic units of living matter.

Alec Jeffreys, of the University of Leicester in Great Britain, invented the technique of "DNA 'fingerprinting,'" as he called it, in 1985. Based on the observation that certain stretches of DNA contain short repeated base sequences that vary considerably in length from person to person, Jeffreys's technique was used first in immigration and paternity cases. It began to be used for identification of criminals in 1987.

Ted Kennedy, Democratic senator from Massachusetts. He cosponsored (with Dianne Feinstein) a Senate bill to ban human cloning that was introduced in early 1998. Biotechnology industry spokespeople and many scientists preferred this bill to a Republican one with the same purpose because, unlike the Republican bill, the Feinstein-Kennedy bill did not ban research on cloned human embryos.

Kenneth Kidd, a Yale University population geneticist. He testified for the prosecution in the pretrial evidence hearing in *United States v. Yee* in the summer of 1990. With Ranajit Chakraborty, he coauthored an article in the December 20, 1991, issue of *Science* that opposed the contention of Richard Lewontin and Daniel Hartl that the usual method of calculating the probability of DNA test samples from two people matching by chance was inaccurate.

Eric Lander, a molecular geneticist at the Whitehead Institute in Cambridge, Massachusetts, testified in *People v. Castro* in 1989 that both the testing procedure used on the DNA evidence in the case and the genetic assumptions used to determine the probability of an accidental match were "scientifically unacceptable." Lander was also on the National Research Council committee that produced a controversial report about DNA profiling in 1992.

Richard Lewontin, a Harvard population geneticist. He stirred up controversy when he and Daniel Hartl claimed, both in the 1990 *United States v. Yee* pretrial hearing and in an article in the December 20, 1991, issue of *Science*, that the standard method of determining the probability that DNA profiles from two people would match by chance was flawed. The statistical calculations, Lewontin said, did not take into account the fact that large categories such as "Caucasians" are really made up of many ethnic subgroups, and people tend to marry within their own subgroup.

Steven Lindow, a plant pathologist at the University of California at Berkeley. He oversaw the first release of genetically altered organisms (bacteria) into the environment in 1987 after years of struggle to win approval from the Recombinant DNA Advisory Committee. His "ice-minus" bacteria protected crops against frost damage.

Victor McKusick, director of medical genetics at Johns Hopkins Medical School in Baltimore, called "the father of genetic medicine." He chaired the National Research Council (NRC) committee that produced a controversial report on forensic DNA testing in 1992.

John C. Mayfield III, a 21-year-old Lance Corporal in the Marines in 1995. He and Corporal Joseph Vlacovsky refused to give samples for DNA testing and archiving, which had been required of all United States military personnel since 1992. Mayfield and Vlakovsky said that the DNA database, which is supposed to help identify the remains of soldiers killed in action, violated their right to privacy. A court-martial in April 1996 convicted the two of refusing to obey a direct order.

Gregor Mendel, an Austrian monk. He bred pea plants in his monastery garden to work out the basic rules by which characteristics are inherited. He published a paper describing his work in 1866, but it remained virtually unknown until several scientists rediscovered it in 1900. Mendel's work became the foundation of genetics and provided the mechanism for evolution by natural selection, first described by Mendel's contemporary Charles Darwin.

John Moore, a Seattle businessman. He sued his doctor (David Golde), the University of California, and several drug companies in 1984 for having made a profitable laboratory cell line from his spleen, which had been removed as a cancer treatment, without consulting him. A lower court supported him, but in 1990 the California Supreme Court ruled that Moore had no ownership claims on his cells once they had been removed from his body.

Thomas Hunt Morgan, pioneer geneticist, worked at Columbia University in New York (1904–28) and later at the California Institute of Technology (1928–45). At Columbia, he and his coworkers and students drew on experiments with fruit flies to relate changes in chromosomes to the inheritance of particular characteristics, thus providing a physical basis for the patterns of trait transmission that Gregor Mendel had observed. Morgan won a Nobel Prize for his work in 1933.

Kary Mullis, discoverer of the polymerase chain reaction (PCR) in 1983, when he was working for Cetus Corporation in Emeryville, California. Using an enzyme called polymerase, PCR repeatedly duplicates DNA, allowing tiny samples to be increased rapidly to a quantity suitable for analysis. It has proved useful in everything from genome sequencing to testing of DNA samples from crime scenes. Mullis won a Nobel Prize for his work in 1994.

Peter Neufeld, a New York defense attorney. He and his partner, Barry Scheck, have become famous for acting in criminal cases involving identification by DNA testing. They founded the Innocence Project, which has used DNA testing to show that certain convicted criminals were actually not guilty. The pair have appeared in a number of famous trials, including the O. J. Simpson murder trial in 1995.

Marya Norman-Bloodsaw, a worker at Lawrence Berkeley Laboratory, a California research facility run by the Department of Energy. In 1995, she and some of her coworkers found out that the laboratory had run several tests, including a genetic test, on samples of their blood and urine without their knowledge or consent. They sued for violation of their rights under the Fourth Amendment and the 1964 Civil Rights Act, and the U.S. Court of Appeals for the 9th Circuit upheld them.

Louis Pasteur, famed French chemist. He provided a scientific basis for part of traditional biotechnology when he showed, starting in the 1850s, that fermentation processes such as those used for millennia to produce wine, beer, buttermilk, and cheese depended on living microorganisms.

Colin Pitchfork, a 27-year-old baker in a village in Leicestershire, England, the first person to be convicted of a crime primarily on the basis of DNA "fingerprinting." When police asked for an unprecedented mass "blooding" (blood sample donation for DNA testing) of young men in several villages in an attempt to solve the rape and murder of two teenaged girls, Pitchfork tried to evade the process by having a coworker give blood in his stead. He was caught, however, and his sample proved to match the semen on one of the victims. He was found guilty of the crimes in 1987.

Robert Pollack, a geneticist. He was perhaps the first to express concern about the safety of recombinant DNA experiments. Then working at Cold Spring Harbor Laboratory in New York, Pollack learned in 1971 that Paul Berg of Stanford was planning to insert a gene from a cancer-causing virus into a type of bacterium that could infect human beings. Pollack persuaded Berg that this might be dangerous, and Berg agreed to transfer the gene between viruses but not go on to infect the bacteria.

Jeremy Rifkin, founder and president of the Foundation on Economic Trends. He has been known as a critic of biotechnology and genetic engineering since the mid-1970s. In addition to filing lawsuits and organizing protest demonstrations, Rifkin has written 14 books on the impact of scientific and technological change on society and the environment. His most recent book is *The Biotech Century* (1998).

Biographical Listing

David Rorvik, a science writer who wrote a best-selling book in 1978 called *In His Image: The Cloning of a Man*, in which he claimed that he had witnessed the cloning of an anonymous wealthy man. Later investigation revealed the book to be a hoax, but it inspired considerable public debate about the ethics of cloning humans.

Steven A. Rosenberg, head of the National Cancer Institute in the National Institutes of Health. A pioneer in devising methods to boost the immune system's ability to fight cancer, in 1989 he became the first person to insert cells containing altered genes into a human being. The genes were not intended to help the man, who had advanced cancer, but only to serve as markers to help Rosenberg track the cells in the man's body. The fact that they caused no harm, however, helped to pave the way for gene therapy.

Barry Scheck, a New York defense attorney. With his partner, Peter Neufeld, he has become famous for acting in criminal cases involving identification by DNA testing. They founded the Innocence Project, which has used DNA testing to show that certain convicted criminals were actually not guilty. The pair have appeared in a number of famous trials, including the O. J. Simpson murder trial in 1995.

Matthias Schleiden, a German biologist, coauthor of the cell theory with Theodore Schwann. In 1839, the two proposed that the microscopic membrane-bound bodies called cells were the units of which all living things were made.

Theodore Schwann, German biologist, coauthor of the cell theory in 1839 with Matthias Schleiden.

Richard Seed, a Chicago physicist. He announced in January 1998 that he was planning to help infertile couples by creating human clones. Most commentators considered Seed eccentric at best and doubted that he had the ability to carry out his plan, but his announcement reignited debates on the ethics of human cloning and calls for legislation to ban such attempts.

Harold T. Shapiro, a Canadian-born economist, president of Princeton University in New Jersey. He is also the chair of the National Bioethics Advisory Commission, which provided a key report to President Bill Clinton in June 1997 on the science and ethics of human cloning. Appropriately enough for this subject, Shapiro is an identical twin—a natural human clone.

Robert Shapiro, CEO of Monsanto Corporation. In 1997 the company spun off the chemical division for which it had been chiefly known and reorganized itself as a "life sciences company" specializing in agriculture, foods, and pharmaceuticals. It owns patents on many important genetically engineered crops. Shapiro maintains that genetically enhanced crops will provide a sustainable way to feed the world's growing

115

population, increasing yield and reducing environmental damage at the same time.

O[renthal]. J. Simpson, a famous African-American former football star. In 1995, he went on trial for the murder of his ex-wife, Nicole Brown Simpson, and an acquaintance of hers, Ronald Goldman. Despite DNA evidence that pointed to his guilt, Simpson was acquitted, probably at least partly because his defense lawyers, who included Barry Scheck and Peter Neufeld, demonstrated that the Los Angeles police could have accidentally or deliberately contaminated evidence from the crime scene with samples of Simpson's blood.

Robert Stillman, a scientist at George Washington University Medical Center in Washington, D.C. With Jerry Hall, he made headlines in October 1993 by successfully cloning early-stage human embryos. Stillman and Hall used embryos that had been scheduled for destruction at a fertility center and did not allow them to develop into fetuses, but opponents of human cloning and of embryo research nonetheless protested their work.

Alfred Sturtevant, a student of Thomas Hunt Morgan, made the first chromosome maps in Morgan's laboratory at Columbia University when Sturtevant was just 19 years old. The maps showed the approximate location of genes coding for certain characteristics on fruit fly chromosomes.

Robert Swanson, a venture capitalist, persuaded Herbert Boyer to join him in exploring the business possibilities of Boyer and Stanley Cohen's newly developed recombinant DNA technology. Swanson and Boyer founded Genentech (from GENetic ENgineering TECHnology) in California in 1976. When the company offered its stock to the public for the first time in 1980, the price per share jumped from $35 to $89 in the first few minutes of trading (a huge increase at the time), even though Genentech had yet to produce a product. Swanson died in December 1999.

Edward L. Tatum, a Stanford University molecular biologist, showed in 1941 that each gene in a DNA molecule coded for one enzyme (that is, a protein molecule). He and his colleague, George Beadle, won a Nobel Prize for this work in 1958.

Harold Varmus, a pioneer researcher on the genetics of cancer, shared a Nobel Prize with J. Michael Bishop, his co-researcher at the University of California, San Francisco, in 1989. He was head of the National Institutes of Health in Bethesda, Maryland, from 1993 to 1999. His positions on biotechnology and genetic engineering issues included a defense of research on embryonic stem cells, which uses cloned human embryos that are not allowed to develop.

Craig Venter, a gene sequencing expert. While working at the National Institutes of Health in the late 1980s, he discovered a method for using short sequences of DNA as "molecular bait" to fish out whole genes from other DNA. The NIH subsequently tried to patent several thousand of these short sequences, called expressed sequence tags. Venter left the NIH in 1992 and founded both a nonprofit organization, the Institute for Genomic Research (TIGR), and a commercial company. In the late 1990s he was working with Perkin-Elmer Corporation, a maker of gene sequencing machines, in a private venture to sequence the human genome.

Joseph Vlakovsky, a 25-year-old Marine corporal in 1995, joined Lance Corporal John C. Mayfield III in refusing to give samples of their DNA for testing and archiving, as had been required of all United States military personnel since 1992. They said that keeping their DNA in a database violated their right to privacy. A court-martial in April 1996 convicted the two of refusing to obey a direct order.

James Dewey Watson, a young American studying at Cambridge University, worked with British molecular biologist Francis Crick to discover the structure of DNA in 1953. For this groundbreaking work, which ultimately showed how inherited information was coded and passed on, Watson, Crick, and Maurice Wilkins received a Nobel Prize in 1962. Watson went on to direct Cold Spring Harbor Laboratory in New York and was also the first head of the Human Genome Project. He resigned in 1992 because of a disagreement about whether NIH should patent a large group of human DNA sequences.

Nancy Wexler, turned personal tragedy into triumph by spearheading the research that discovered the gene that causes Huntington's disease, the inherited brain disorder that killed her mother and which she herself has a 50 percent chance of inheriting. She was also the first head of the Human Genome's Ethical, Legal, and Social Issues (ELSI) working group. She teaches at Columbia University in New York.

Ian Wilmut, a scientist at the Roslin Institute in Scotland. He startled the world in February 1997 by announcing that he and his coworkers had cloned a sheep (which they named Dolly) from a mature body cell, something many scientists had thought impossible. Wilmut was looking for a technique to produce herds of identical, genetically engineered animals that could produce drugs and other substances, but his discovery spurred debates on human cloning as well.

CHAPTER 5

GLOSSARY

Biotechnology, genetic engineering, and genetics are complex fields with highly technical vocabularies. This chapter presents some of the terms that the general reader is most likely to encounter while researching these fields and their ethical, legal, and social implications. Several web sites also offer online glossaries (see Chapter 6, "How to Research Biotechnology and Genetic Engineering").

ADA (adenosine deaminase) deficiency   A rare inherited disorder in which lack of a certain enzyme in white blood cells causes an essentially complete failure of the immune system. It was the first illness to be treated successfuly by gene therapy (in 1990).

adenine   One of the four bases in DNA and RNA. It pairs with thymine in DNA and uracil in RNA.

Agrobacterium tumefaciens   A bacterium that infects plant cells and injects its DNA into them. It was the first vector that scientists used to insert foreign genes into plants, and it is still widely used in agricultural biotechnology.

allele   An alternate form of a gene or stretch of DNA (locus). Different alleles may produce different traits; for instance, one allele at a locus determining eye color might produce blue eyes, while a different allele at that locus might produce brown. Other differences in alleles produce no visible effect but are useful in DNA identification testing and in tracing inheritance patterns of traits (such as those related to disease) in families.

Americans with Disabilities Act (ADA)   Passed in 1990, this act protects the disabled from discrimination. It covers healthy people who are perceived as disabled, which may include those with inherited defects revealed by genetic tests.

amino acids   The small molecules of which proteins are made. There are 20 different kinds.

Glossary

Animal and Plant Health Inspection Service (APHIS)   The part of the U.S. Department of Agriculture responsible for regulating genetically altered agricultural plants and animals.

Asilomar conference   A conference held at a California retreat center in February 1975, in which 140 geneticists and molecular biologists worked out safety standards for recombinant DNA experiments.

autoradiograph   The picture produced by exposing a membrane containing a DNA sample and radioactively labeled probes to X-ray film. It shows a series of bands something like a supermarket bar code. Autoradiographs from different samples are compared in DNA identification testing.

***Bacillus thuringensis* (Bt)**   A bacterium that produces a toxin that kills a variety of pest insects but is harmless to most nonpest insects and other living things, including humans. Sprays containing the bacteria have been used as short-lived, "organic" insecticides. The gene that produces the Bt toxin has recently been inserted into crop plants.

bandshifting   A problem in comparing DNA profile samples that arises when the DNA from one sample migrates through an electrophoresis gel at a speed different from that of another sample, displacing all the bands relative to each other. Including a band that can be expected to match in both samples helps to correct for bandshifting.

bases   Small molecules that usually exist in pairs in DNA and RNA. In DNA, the bases are adenine, thymine, guanine, and cytosine; in RNA, uracil replaces thymine. The sequence of bases in a nucleic acid molecule contains the genetic code.

Bayh-Dole Act   This act, passed in 1980, encourages technology transfer by allowing universities to take out patents on discoveries they make with government funding.

biochip   See **DNA chip**.

bioinformatics   The application of computer and statistical techniques, such as database management, to the organization of biological information, such as DNA or protein sequences.

bioprospecting   Searching an environment for living things or parts of living things (including genetic material) that may have commercial use. Critics call this process "biopiracy."

bioremediation   A branch of biotechnology that uses living things, usually microorganisms, to repair environmental damage, for instance by breaking down oil or other toxic chemicals.

biotechnology   Using or altering living things in processes that benefit humankind; now frequently applied to commercial processes that use organisms with altered genes.

bovine growth hormone (bovine somatotropin, BGH, BST)   A cattle hormone that can be made by bacteria containing recombinant DNA (the recombinant hormone is known as rBGH). The hormone is given to dairy cattle to increase milk production. Its use is controversial.

BRCA1, BRCA2   Genes found in the early 1990s to be inherited in some families in which breast and ovarian cancer are unusually common and occur at an early age. Five to 10 percent of women with breast cancer have a mutated form of one of these genes, which can be detected by a test.

Bt   See *Bacillus thuringensis*.

carrier   An organism that has inherited a gene related to a disease and therefore can pass it on to offspring but does not suffer from the disease.

ceiling principle   A method for calculating the probability of a chance match between DNA samples from two different people, suggested in a 1992 National Research Council report as a compromise to resolve a dispute about the statistics cited in forensic DNA testing.

cell   The basic unit of which all organisms are composed, made up of a microscopic piece of living material, surrounded by a membrane. It is the simplest living system that can exist independently.

cell line   A group of cells altered so that they will multiply indefinitely in culture in a laboratory.

chimera   A transgenic animal. The original chimera was a monster in Greek mythology that was part lion, part goat, and part dragon.

chromosome   One of a group of threadlike bodies containing the DNA that carries a cell's basic genetic information (they also contain protein). In cells with nuclei, they are found in pairs inside the nucleus. Chromosomes ("colored bodies") reproduce themselves during cell division.

clone   A gene, cell, or organism that is the exact genetic duplicate of another gene, cell, or organism; both are produced asexually from the same ancestor. Genes or other stretches of DNA may be cloned in bacteria for study or for biotechnology processes. Plant and (to a more limited extent) animal clones are used in biotechnology and research. The ethics of cloning humans has been hotly debated.

Combined DNA Index System (CODIS)   A national DNA database established by the FBI in October 1998 to coordinate information from DNA samples taken from criminals by the states.

cytoplasm   The living material in the main body of a cell (outside the nucleus in nucleated cells).

cytosine   One of the four bases in DNA and RNA. It always pairs with guanine.

deoxyribose   The sugar that, with phosphate, makes up the "backbones" in a DNA molecule.

differentiation   The process of maturation in which a cell takes on characteristics associated with a particular type of tissue, such as nerve tissue or muscle tissue. Differentiation is usually, but not always, irreversible.

Glossary

DNA (deoxyribonucleic acid)   The chemical in which inherited information is encoded, except in some viruses. Each DNA molecule consists of two phosphate-sugar backbones twined around each other in the shape of a double helix. Pairs of bases (adenine, thymine, guanine, and cytosine) joined by weak hydrogen bonds stretch between the backbones like rungs on a twisted ladder.

DNA chip (gene chip, biochip)   A device, invented by Affymetrix in 1996, that consists of a computer chip coated with DNA (GeneChip is Affymetrix's trademarked term). It allows thousands of genes to be scanned for mutations at a time.

DNA "fingerprinting" (profiling, identification testing)   A technique, invented by Alec Jeffreys of the University of Leicester in Great Britain in the early 1980s, that uses stretches of DNA that differ considerably from person to person as a means of determining whether a sample of DNA came from a particular person. It is most often used to determine family relationships or to find out whether a sample of DNA from a crime scene could have come from a certain suspect.

DNA Identification Act   Passed in 1994, this law gives the FBI responsibility for overseeing training, funding, and proficiency testing of all laboratories doing forensic DNA profiling.

DNA probe   A sequence of single-stranded DNA, labeled for detection in some way (such as with radioactivity), that is used to bind to and mark other single-stranded DNA with a complementary base sequence (that is, with a sequence opposite to that of the first one). DNA probes are used in genome mapping and DNA identification testing.

Dolly   A sheep (named after Dolly Parton) cloned from an udder cell of an adult ewe by Ian Wilmut and his colleagues at the Roslin Institute in Scotland. She is the first mammal to be cloned from a mature adult cell. Her existence was announced in an article in *Nature* on February 27, 1997.

dominant gene   A gene that is expressed even if only one copy of it is inherited. Compare with **recessive gene**.

double helix   The shape of a DNA molecule, in which two parallel strands twist or coil like a corkscrew.

EcoR1   The restriction enzyme from *Escherichia coli* bacteria used by Herbert Boyer and Stanley Cohen in their first recombinant DNA experiments.

electrophoresis   A laboratory procedure used to separate large molecules, such as fragments of nucleic acid or protein, from a mixture of similar molecules. When an electric current is passed through the mixture, it causes the molecules to travel through a medium (such as a gel) at speeds that depend on their molecular weight and electrical charge. The molecules are thus separated into a series of bands or spots in the medium.

enzyme   A protein that catalyzes a chemical reaction, speeding the rate at which it occurs.

Escherichia coli (E. coli)   A common type of bacterium that usually lives harmlessly in the human intestine (though some strains of it can cause serious illness). It grows easily in the laboratory and was frequently used in early recombinant DNA and other genetics experiments.

eugenics   A pseudoscience founded in the late 19th century by Francis Galton (he coined the word, from Greek words meaning "well born," in 1883). It holds that complex behaviors and characteristics such as intelligence are inherited and that the human race can be improved by encouraging people with desirable traits to reproduce and discouraging or forcibly preventing those with undesirable traits from doing so.

eukaryote   An organism made up of one or more cells with nuclei. All living things except viruses, bacteria, and blue-green algae are eukaryotes. Compare with **prokaryote**.

exon   The part of a gene that carries the code for making a protein. Compare with **intron**.

expressed sequence tag (EST)   The portion of a gene expressed by messenger RNA in the making of a protein. ESTs are the only parts of a cell's DNA that are actually active in that cell, so they are likely to code for important compounds. ESTs can be used as probes to find whole genes.

Federal Food, Drug, and Cosmetic Act (FFDCA)   First passed in 1938, this act gave the FDA and, later, the EPA the right to set tolerances for certain substances on and in food and feed. Herbicide residues on genetically engineered herbicide-tolerant food crops and pesticides (such as Bt toxin) produced by genetically engineered food crops are regulated under the FFDCA.

Federal Insecticide, Fungicide, and Rodenticide Act (FIFRA)   A law, passed in 1947, that governs the testing and use of pesticides. After the Environmental Protection Agency (EPA) was established in 1970, it took over the administration of regulations required by FIFRA. FIFRA has been amended to include genetically engineered crops that produce Bt (*Bacillus thuringensis*) toxin in its definition of pesticides.

fermentation   A group of processes in which microorganisms such as bacteria or yeast cause chemical changes that include the production of gas. Fermentation is part of many traditional biotechnology processes, such as those used to make alcoholic beverages. It was thought to be strictly a chemical reaction until the mid-19th century, when Louis Pasteur proved that it required living microorganisms.

Flavr Savr   A type of tomato that a company called Calgene genetically altered to delay its breakdown after ripening. In 1994, Flavr Savr became the first genetically modified food to go on sale.

Glossary

Frye rule A rule (first stated in a 1923 case, *Frye v. United States*) that judges often use to decide whether evidence based on new scientific techniques will be admitted in court. The rule recommends accepting such evidence if the technique on which it is based has "gained general acceptance" among scientists in its field.

functional food A food claimed to improve the body's performance or prevent disease.

gene The basic unit of inherited information, consisting of a sequence of nucleotides in a DNA molecule that carries the code for production of a specific protein or RNA molecule.

gene chip See **DNA chip.**

gene splicing Common term for insertion of one or more genes from one species into the genome of another; synonym for recombinant DNA technology.

gene therapy Treatment of a disease by altering genes. Gene therapy was first successfully used on a human being in September 1990.

Genentech The first biotechnology company to be based on recombinant DNA technology. Herbert Boyer and Robert Swanson founded it in 1976.

genetic code The sequence of bases or nucleotides in a DNA or RNA molecule that determines the sequence of amino acids in a protein. Each "letter" of the code consists of a group of three bases or nucleotides.

genetic determinism The belief that most physical and mental characteristics, including complex behaviors, are determined primarily or exclusively by genetics (heredity).

Genetic Diseases Act A law, passed in 1976, that establishes screening to identify carriers of inherited diseases but specifies that participation in such screening is to be voluntary and not linked to eligibility for federal services.

genetic engineering Direct manipulation of genetic information or transfer of genes from one type of organism to another to produce new biological structures or functions useful to human beings. Genetic engineering includes, but is not limited to, recombination of DNA.

genetic screening Testing of a population to identify people at risk for suffering from a genetic disease or passing such a disease to their children.

genetics The study of the patterns and mechanisms by which traits are inherited.

genome An organism's complete collection of genetic information.

genomics The science of identifying genes and their functions, including building maps and databases of genes.

genotype The nature of an organism or group as determined by its genes. Compare with **phenotype**.

germ-line genes Genes that are contained in the sex cells (the cells that will become sperm and eggs) and therefore can be passed on to offspring.

GM (genetically modified) foods   European term for foods that contain, or may contain, transgenic organisms.

guanine   One of the four bases in DNA and RNA. It always pairs with cytosine.

Harvard Oncomouse   A type of mouse genetically engineered to be unusually susceptible to cancer. Intended for use in medical research and testing of possible carcinogens, it was patented by Harvard in 1988. It was the first genetically altered animal to be patented.

Health Insurance Portability and Accountability Act (HIPAA)   Passed in 1996, this act includes a provision forbidding health insurers that issue group plans to employers from denying insurance on the basis of preexisting genetic conditions. It does not apply to individual plans or to employers that ensure their own workers.

hemoglobin   The iron-containing pigment in red blood cells that gives the cells (and the blood) their color and carries oxygen throughout the body. Several inherited blood diseases, including sickle-cell disease, produce abnormal forms of hemoglobin.

HTLV-1 (Human T-cell Leukemia Virus 1)   A virus that causes a rare blood cancer in humans; it is related to the virus that causes AIDS.

Human Genome Project   An international project, launched in 1990, that aims to sequence all of the approximately 100,000 genes in the human genome. As of mid 1999, project leaders expected to complete that task by 2002.

Human Genome Diversity Project   Established in 1992 and headed by Stanford geneticist Luca Cavalli-Sforza, this project aims to collect DNA samples from about 500 of the world's most endangered indigenous peoples and analyze them to determine such things as past human migration patterns and differences in susceptibility to disease. Critics call it the "vampire project" and say it will exploit native peoples' genetic heritage.

Huntington's disease   An inherited form of incurable brain degeneration caused by a single dominant gene. Affected people usually show no symptoms until middle age. A person who inherits even one copy of the gene will develop the disease, and the child of someone with the disease has a 50 percent chance of developing it. The gene that causes the disease was identified in 1993 and can be detected by a test.

hybrid   An offspring of parents that differ in variety or species.

ice-minus bacteria   A genetically altered form of *Pseudomonas syringae* bacteria, developed by Steven Lindow and associates at the University of California, Berkeley, in the early 1980s. These bacteria lack a gene that allows the normal form of the bacteria to produce ice on plants; they therefore were expected to protect plants against frost damage. In 1987, ice-minus bacteria became the first genetically altered organisms to be released into the environment.

insulinlike growth factor 1 (IGF-1)   A hormone, related to bovine growth hormone (BGH), that is thought to increase the risk of breast and prostate cancer in humans. Some people fear that drinkers of milk from cows given recombinant bovine growth hormone will consume small amounts of IGF-1 and thus increase their risk of cancer.

intron   A DNA base sequence that interrupts the protein-coding portion of a gene (exon). The function of introns is currently unknown, so they are sometimes called "junk DNA."

in vitro ("in glass") fertilization   Combination of an egg and a sperm in a laboratory to create an embryo that is then implanted in a uterus; also called test-tube fertilization.

lambda   A type of virus that infects bacteria. It was used in the first recombinant DNA experiments.

ligase   A type of enzyme that permanently links complementary single strands of DNA.

locus (pl. *loci*)   The place on a chromosome occupied by a gene or other specified sequence of DNA used as a marker.

lymphokine   One of a family of immune system chemicals that stimulate the activity of immune cells. Interferon is an example.

meiosis   The form of cell division that produces sex cells. Unlike standard cell division, meiosis results in cells that have single chromosomes rather than pairs and thus contain only half the normal number of genes. Compare with **mitosis**.

messenger RNA   The form of RNA that is transcribed (copied) from DNA and moves from the cell nucleus into the cytoplasm, where it acts as a template for the formation of proteins.

minisatellite   The type of variable DNA sequence that British geneticist Alec Jeffreys used as the basis for his "'DNA 'fingerprinting'" identification test.

mitochondrial DNA   DNA contained within mitochondria, organelles that help a cell use energy. Unlike nuclear DNA, mitochondrial DNA is inherited only through the female line. It has been used to trace family relationships.

mitosis   The process by which cells (except those that form sex cells) divide. It produces two daughter cells that are genetically identical to each other and to the parent cell. Compare **meiosis**.

molecular biology   A branch of biology dealing with the molecular basis of biological processes such as protein synthesis and transmission of inherited characteristics.

monoculture   The practice of planting large areas with the same kind of crop.

Biotechnology and Genetic Engineering

Monsanto Corporation A multinational corporation, headquartered in St. Louis, Missouri, that was formerly known chiefly as a chemical company. In 1997 it reorganized itself as a "life sciences company," specializing in agriculture, food, and pharmaceuticals. It owns patents on many genetically engineered crops.

moratorium A temporary halt to an activity.

multiplication rule A rule used to determine the probability of DNA test samples from two people (other than identical twins) matching by chance. It states that if the variations (alleles) at the tested loci are inherited independently of each other, the probability of a chance match within a certain large population can be determined by multiplying the probabilities for inheritance of the individual alleles in that population. In the early 1990s, some population geneticists questioned the accuracy of this rule because they doubted the validity of the assumptions on which it is based.

mutation An inheritable change in a DNA sequence.

National Bioethics Advisory Commission (NBAC) A presidential advisory commission consisting of 18 experts in theology, science, and law and headed by Princeton University president Harold Shapiro. It has issued reports on such subjects as human cloning.

National Institutes of Health (NIH) A group of large, prestigious medical research institutions funded by the United States government and located in Bethesda, Maryland.

natural selection The mechanism of evolution described by Charles Darwin in *On The Origin of Species* (1859) and independently by Alfred Russel Wallace. It states that the members of a species with inherited characteristics that make them most suited to survive in a particular environment will be most likely to survive and bear young. Over generations, therefore, a change in the environment can cause a change in the predominant characteristics of a species.

nuclear transfer A technique for cloning in which the nucleus of a cell (embryonic or, more recently, mature) is transferred into an unfertilized egg from which the nucleus has been removed. The two are then fused by electricity. The resulting single cell develops into a clone of the organism that provided the nucleus.

nucleic acid A large molecule composed of nucleotides. The most common nucleic acids are DNA and RNA.

nucleotide A subunit of DNA or RNA composed of a base and an attached "backbone" of phosphate and sugar. Each DNA or RNA molecule contains thousands of nucleotides linked together.

nucleus The membrane-bound central body in eukaryotic cells that contains the cell's main genetic material.

Glossary

P1–P4 Levels in a safety classification system for recombinant DNA experiments with microorganisms (and research on microorganisms in general), first used in the mid-1970s. P1 was the lowest danger level, P4 the highest.

PCR See **polymerase chain reaction.**

pharming Use of genetically altered farm animals to produce human body chemicals and drugs.

phenotype An organism's physical appearance. The phenotype may or may not match the genotype. Compare with **genotype.**

Plant Patent Act Passed in 1930, this act stops short of allowing conventional patents on plants but allows plant breeders to forbid others to clone (reproduce asexually) hybrid varieties that they have developed.

Plant Pest Act Passed in 1987, this act gives the U.S. Department of Agriculture's Animal and Plant Health Inspection Service (APHIS) the right to regulate any organisms, including those produced by genetic engineering, that are or might be plant pests.

Plant Variety Protection Act This act, passed in 1970, extends the Plant Patent Act to forbid sexual reproduction of plant varieties without their developers' permission.

plasmid A circular DNA molecule, used by some bacteria to transfer genes from one microorganism to another. It is separate from the main bacterial genome and can reproduce on its own. Herbert Boyer and Stanley Cohen used plasmids in some of the first recombinant DNA experiments.

polymerase chain reaction (PCR) A method for rapidly multiplying copies of a DNA sequence, developed in the early 1980s by Kary Mullis. It can be used to increase tiny samples of DNA to a size usable in DNA identification testing or gene sequencing.

prokaryote A cell or organism that lacks a separate nucleus, such as a bacterium. Compare with **eukaryote.**

prosecutor's fallacy A confusion between the two questions to be asked about DNA identification evidence in a court case: (1) What is the probability that an individual's DNA will match that found at a crime scene, given that the person is innocent? and (2) What is the probability that an individual is innocent, given that there is a match? The fallacy lies in giving the answer to the first question in response to the second.

protein One of a large family of substances that are composed of amino acids arranged in a certain order. Proteins carry out most functions in cells, including acting as enzymes, structural components, and signaling molecules. They are assembled according to instructions in the genetic code contained in DNA and RNA. Instructions for each type of protein are carried on a separate gene.

Pseudomonas syringae A common type of bacterium that breaks up plant cells by encouraging ice crystals to form on them and then feeds on the remains. In 1987, *Pseudomonas* bacteria altered to make them incapable of causing ice formation ("ice-minus") became the first genetically engineered organisms deliberately released into the environment.

rBGH (recombinant bovine growth hormone) See **bovine growth hormone**.

reach-through rights Rights demanded by some holders of biotechnology patents as part of licensing agreements. If a company has reach-through rights to a technology, it can demand royalties on all products developed with that technology, even if the development work is done by others.

recessive gene A gene that does not produce a detectable characteristic unless copies of that gene have been inherited from both parents. Compare with **dominant gene**.

recombinant DNA DNA from different types of organisms that is combined directly in a laboratory.

Recombinant DNA Advisory Committee (RAC) A group of experts formed by the National Institutes of Health in October 1974 to judge the safety of recombinant DNA experiments. It still exists, but it now reviews only experiments that differ from previous ones in a substantial way.

regulatory gene A gene whose job is to modify the action of other genes rather than to produce a protein. Compare with **structural gene**.

restriction enzyme (restriction endonuclease) An enzyme produced by some bacteria to break up the DNA of invading viruses. A restriction enzyme cuts a DNA molecule at any spot where a particular sequence of bases occurs, leaving a piece of single-stranded DNA at each end. There are many types of restriction enzymes, each of which cuts at a different sequence. Restriction enzymes have been used in recombinant DNA technology since its start.

restriction fragment length polymorphism (RFLP) A variation in DNA sequence that produces fragments of different lengths when the same restriction enzyme cuts DNA from different people. RFLPs have been used as markers in genome mapping, identifying genes associated with disease, and DNA identification testing.

RNA (ribonucleic acid) A single-stranded nucleic acid with a molecular structure similar to that of DNA but with uracil substituting for thymine among the bases. Unlike DNA, RNA is found in the cytoplasm of the cell as well as the nucleus. It plays a vital role in turning the DNA code into protein molecules and in other chemical activities of the cell. It exists in several forms, including messenger RNA and transfer RNA.

"Roundup Ready" crops Crops genetically engineered by Monsanto Corporation to be resistant to glyphosate, a type of herbicide that the company markets under the name Roundup.

sex cells   The cells that become eggs and sperm and carry inherited material to the next generation. Created by meiosis, each sex cell contains half the usual complement of chromosomes. Thus, when an egg and a sperm cell unite in fertilization, the resulting fertilized egg will have the normal number of chromosomes.

short tandem repeats (STRs)   Areas of DNA containing many short, identical sequences in a row. The number of repeats varies considerably from person to person. In the late 1990s, a test using STRs has become the most commonly used form of DNA profiling.

Sickle-Cell Anemia Act   This act, passed in 1972, established a national screening program to identify and counsel carriers of sickle-cell anemia. It specified that participation in the screening was to be voluntary and not linked to eligibility for federal services.

sickle-cell disease (sickle-cell anemia)   An inherited blood disease, fairly common among people of African descent, that is caused by abnormal hemoglobin produced by a defective gene. The mutant hemoglobin deforms the round cells to a sickle shape and causes them to block tiny blood vessels, starving the body of oxygen and causing pain, illness, and sometimes an early death. Because the gene that causes the disease is recessive, people develop it only if they inherit the mutant gene from both parents. A person with only one mutant gene is called a sickle-cell carrier or a possessor of sickle-cell trait.

sickle-cell trait   A trait possessed by people, usually of African descent, who inherit one normal and one mutant copy of the gene that, when a person inherits two copies, causes sickle-cell anemia. People with sickle-cell trait are also called sickle-cell carriers because they can pass the defective gene to their children. They themselves, however, are perfectly healthy. Indeed, they seem to have unusual resistance to malaria, a serious blood disease caused by a parasite that is widespread in Africa. People with sickle-cell trait have sometimes been discriminated against in employment or insurance because of the mistaken impression that they are or will become ill.

somatic cell   Any cell in the body except the sex cells and their precursors.

somatic cell nuclear transfer   A cloning technique in which the cloned cell is a mature body cell. This technique was used to create Dolly the sheep. See also **nuclear transfer**.

stem cell   A cell that has not yet differentiated and has the potential to produce cells of a wide range of types, in some cases any type. The study of stem cells may eventually produce tissues for transplantation or treatment of degenerative diseases.

"sticky ends"   Pieces of single-stranded DNA at the ends of a double-stranded segment left after cutting by restriction enzymes or certain other treatments. They attract and bind to other pieces of single-stranded DNA with a complementary sequence. This fact was used in the creation of recombinant DNA in the early 1970s.

structural gene   A gene that contains coded instructions for making a particular protein or RNA molecule. Compare with **regulatory gene**.

sustainable agriculture   Agriculture that does not damage or deplete the environment.

SV40 (Simian Virus 40)   A monkey virus that can cause cancer in mice. Paul Berg used it in the first recombinant DNA experiments.

technology transfer The process of converting the results of basic scientific research into commercially useful products.

thalassemia   An inherited blood disease usually found in people of Mediterranean descent. It was the target for Martin Cline's failed attempt at gene therapy in 1980.

thymine   One of the four bases in DNA. It always pairs with adenine. In RNA it is replaced by uracil.

Toxic Substances Control Act (TSCA)   An act passed in 1976 that gives the Environmental Protection Agency the right to regulate release and sale of "new" chemicals. TSCA was amended in 1986 to classify genetically engineered organisms expressing new traits or containing genes from two different genera as equivalent to new chemicals and thus susceptible to regulation under the act.

transfer RNA   A type of RNA that bonds to molecules of particular amino acids and brings them to the site where they are assembled, in the order specified by messenger RNA, to form a protein.

transgenic   Containing genes from two or more species.

uracil   One of the four bases in RNA. It is the equivalent of thymine in DNA and pairs with adenine.

variable number of tandem repeats (VNTR)   A stretch of DNA containing repeated short base sequences that vary in number from one person to another. A test for VNTR loci, a form of RFLP testing, was the more accurate of the two DNA identification test methods used in the early 1990s, but it was time consuming and required a relatively large sample.

vector   A tool for carrying genes from one organism into the genome of another. Plasmids and certain viruses are examples of vectors.

X-ray crystallography   A method of visualizing the arrangement of atoms in a molecule by shining an X-ray beam through a crystal of the test substance and exposing the refracted beam to photographic film.

PART II

GUIDE TO FURTHER RESEARCH

CHAPTER 6

HOW TO RESEARCH BIOTECHNOLOGY AND GENETIC ENGINEERING

The tremendous growth in the resources and services available through the Internet (and particularly the World Wide Web) is providing powerful new tools for researchers. Mastery of a few basic online techniques enables today's researcher to accomplish in a few minutes what used to require hours in the library poring through card catalogs, bound indexes, and printed or micro-filmed periodicals.

Not everything is to be found on the Internet, of course. While a few books are available in electronic versions, most must still be obtained as printed text. Many periodical articles, particularly those more than ten years old, must still be obtained in hard copy form from libraries. Nevertheless, the Internet has now reached "critical mass" in the scope, variety, and quali-ty of material available. Thus, it makes sense to make the Net the starting point for most research projects. This is particularly true regarding recent events in biotechnology and genetic engineering. Web/Internet links can lead the researcher not only to companies, professional organizations, and journals in the field but even to complex databases of DNA sequences and other highly technical material.

THINKING LIKE A SPIDER: A PHILOSOPHY OF WEB SEARCHING

For someone not used to it, searching the Internet (the Net) and the World Wide Web can feel like spending hours trapped inside a pinball machine. The shortest distance between a researcher and what he or she wants to know is seldom a straight line, at least not a single straight line. These things are called nets and webs for good reason: Everything is connected by links, and often a researcher must travel through a number of these to find the desired information.

Net/web searching is best approached with a combination of patience, alertness, and, preferably, humor. A given search often will not reveal the desired information but will unearth at least three things, or groups of things, that are even more interesting. The information sought on the initial search, meanwhile, will be uncovered by chance at a later time when the researcher is looking for something else entirely. The sooner one accepts this, the sooner searching is likely to become rewarding rather than painful. In addition to specific files related to particular areas of research, it is a good idea to maintain a general file into which promising URLs (universal resource locators or web addresses) or pieces of web sites can be copied as they are encountered.

It is easy to feel lost on the web, but it is also easy to find one's way back. During any given search, the Back button on the browser is the Ariadne's thread that will guide the researcher back through the labyrinth to the beginning of the adventure on the browser's home page, passing en route through all the sites visited (so that one can stop for another look or, if desired, jump off to somewhere else). The History button provides a list of all the sites visited on recent previous sessions.

Finally, a word of caution about the Internet. It is important to critically evaluate all materials found on the Net. Many sites have been established by well-known, reputable organizations or indviduals. Others may come from unkown individuals or groups. Their material may be equally valuable, but it should be checked against reliable sources. Gradually, each researcher will develop a feel for the quality of information sources as well as a trusty toolkit of techniques for dealing with them.

TOOLS FOR ORGANIZING RESEARCH

Several techniques and tools can help the researcher keep materials organized and accessible:

Use the web browser's "Favorites" or "Bookmarks" menu to create a folder for each major research topic (and optionally, sub-folders). For example, folders used in researching this book included: organizations, laws, cases, current news, reference materials, and bibliographical sources.

Use favorites or bookmark links rather than downloading a copy of the actual web page or site, which can take up a large amount of both time and disk space. Exception: if the site has material that will definitely be needed in the future, download it to guard against it disappearing from the web.

If a whole site needs to be archived, obtain one of a variety of free or low-cost utility programs such as WebWhacker, which make it easier to download a whole site automatically, including all levels of links. But use the program judiciously: a site such as www.eff.org may contain gigabytes of material. Where applicable, "subscribe" to a site so it will automatically notify you when new material is available.

Use a simple database program (such as Microsoft Works) or, perhaps better, a free-form note-taking program (such as the shareware program WhizFolders, available at http://skanade.simplenet.com/. This makes it easy to take notes (or paste text from web sites) and organize them for later retrieval.

WEB INDEXES

A web index is a site that offers what amounts to a structured, hierarchical outline of subject areas. This enables the researcher to zero in on a particular aspect of a subject and find links to web sites for further exploration.

The best known (and largest) web index is Yahoo! (www.yahoo.com). The home page gives the top-level list of topics. Four of these are of particular use for researching biotechnology and genetic engineering:

Science: Selecting Biology under this heading produces a host of useful subheads, including archives, ask an expert, biochemistry, biodiversity, biological safety, biomedical ethics, biotechnology, cell biology, companies, genetics, institutes, journals, molecular biology, organizations, and pharmacology.

Business and Economy: Selecting Companies under this topic produces several listings related to biotechnology, including agriculture, biomedicine, environment, food, government, health, law, manufacturing, and technology transfer.

Government: Provides listings including documents, ethics, law, and research labs. Categories under the Law subhead that are likely to be useful include cases, disabilities, employment law, environmental, health, indigenous peoples, intellectual property, legal research, and technology.

Health produces subheads dealing with medicine and with specific diseases, including genetic disorders.

In addition to following Yahoo's outline-like structure, there is also a search box into which the researcher can type one or more keywords and receive a list of matching categories and sites.

Web indexes such as Yahoo have two major advantages over undirected surfing. First, the structured hierarchy of topics makes it easy to find a particular topic or subtopic and then explore its links. Second, Yahoo does not make an attempt to compile every possible link on the Internet (a task that is virtually impossible, given the size of the web). Rather, sites are evaluated for usefulness and quality by Yahoo's indexers. This means that the researcher has a better chance of finding more substantial and accurate information. The disadvantage of web indexes is the flip side of their selectivity: The researcher is dependent on the indexer's judgment for determining what sites are worth exploring.

Two other web indexes are LookSmart (http://www.looksmart.com) and the Mining Company's About.com (http://home.miningco.com.)

SEARCH ENGINES

Search engines take a very different approach to finding materials on the web. Instead of organizing topically in a "top down" fashion, search engines work their way "from the bottom up." Basically, a search engine consists of two pieces of software. The first is a "web crawler" that systematically and automatically surfs the Net, following links and compiling them into an index with keywords (drawn either from the text of the sites themselves or from lists of words that have been flagged in a special way by the site's creators). The second program is the search engine's "front end:" It provides a way to match user-specified keywords or phrases with the index and display a list of matching sites.

There are hundreds of search engines, but some of the most widely used include:

AltaVista (http://www.altavista.com)
Excite (http://www.excite.com)
Google (http://www.google.com)
Hotbot (http://www.hotbot.com)
Infoseek (http://www.infoseek.com)
Lycos (http://www.lycos.com)
Magellan (http://www.mckinley.com)
NetFind (AOL) (http://www.aol.com)
Northern Light (http://www.nlsearch.com)
WebCrawler (http://www.WebCrawler.com)

Search engines are generally easy to use by employing the same sorts of keywords that work in library catalogs. There are a variety of web search tutorials available online (try "web search tutorial" in a search engine). One good one is published by The Web Tools Company at http://thewebtools. com/tutorial/tutorial.htm.

Here are a few basic rules for using search engines:

When looking for something specific, use the most specific term or phrase. For example, when looking for information about DNA fingerprinting, use "DNA fingerprinting" or "forensic DNA testing," not "DNA." (When using phrases as search specifications, enclose them in quotation marks.)

When looking for a more general topic, use several descriptive words (nouns are more reliable than verbs), such as privacy genetic information. (Most engines will automatically put pages that match all three terms first on the results list.)

Use "wildcards" when a desired word may have more than one ending. For example, gene* matches genetics, genetic engineering, genome, and so on.

Most search engines support Boolean (AND, OR, NOT) operators, which can be used to broaden or narrow a search.

Use AND to narrow a search. For example, agriculture AND biotechnology will match only pages that have both terms.

Use OR to broaden a search. Agriculture OR biotechnology will match any page that has *either* term.

Use NOT to exclude unwanted results. Biotechnology NOT agriculture finds articles about biotechnology other than agricultural biotechnology.

Since each search engine indexes somewhat differently and offers somewhat different ways of searching, it is a good idea to use several different search engines, especially for a general query. Several "metasearch" programs automate the process of submitting a query to multiple search engines. These include:

Dogpile (http://www.dogpile.com)
Inference FIND (http://www.infind.com)
Metacrawler (http://www.go2net.com/search.html)
SavvySearch (http://www.savvysearch.com)

METASITES: EVERYTHING ABOUT BIOTECH AND THEN SOME

One basic principle of research is to take advantage of the fact that other people may have already found and organized much of the most useful information about a particular topic. For biotechnology and genetic engineering, there are several web sites that can serve as excellent starting points for research because they provide links to vast numbers of other resources.

Biotechnology and Genetic Engineering

- The National Biotechnology Information Facility, http://nbif.org, has a search engine, a news page, and an extensive collection of links to sites related to legal and regulatory aspects of biotechnology, including policies, permit databases, biotechnology standards, and online legal resources. It also provides access to bioinformatics and biotechnology databases.

- Access Excellence, http://www.accessexcellence.org/, is aimed at biology teachers and students. It has news about biotechnology, educational activities, and links to numerous sites related to biotechnology and genetics.

- The Biotechnology Information Institute, at http://www.bioinfo.com/, has extensive links to sites in genetics and biomedicine.

- Deakin University's Biological and Chemical Sciences Department has a good list of links on subjects in biology and chemistry, including FBI DNA testing, the Human Genome Project, and human genome maps, at http://www.deakin.edu.au/fac_st/bcs/goodies/biol.html.

- The Genetic Education Center of the University of Kansas Medical Center, http://www.kumc.edu/gec, has links related to the Human Genome Project, genetic conditions, educational resources, glossaries, and more. Links to numerous national and international professional organizations can be found at http://www.kumc.edu/instruction/medicine/genetics/prof/soclist.html.

- The Institute for Genomic Research (TIGR) provides a links page at http://www.tigr.org/links. The links cover news, organizations, genome maps and sequences, and more.

- The online version of the journal *Nature Biotechnology* has a Web Extras page, http://biotech.nature.com/web_extras, that includes many biotechnology and genetics links, grouped under general information, news sites, databases, government agencies, and industry organizations.

- Indiana University and the University of Texas have a combined site, Biotech: Life Science Resources and Reference Tools, at http://biotech.icmb.utexas.edu. It includes an illustrated dictionary, an extensive list of Science Resources links, and BioMedLink, a mega-database of biomedical sites.

- Cato Research, Ltd., at http://www.cato.com/biotech, provides the World Wide Web Virtual Library for Biotechnology, covering biotech, drug development, genetic engineering, and related fields. At http://www.cato.com/crl/links.html, it provides links to government resources and to the biotechnology industry, both in the United States and in Britain.

- Muritech, at http://www.muritech.com/mp/biosites.html, has a list of web links that includes multimedia sites and a variety of electronic journals.

SPECIFIC INTERNET/WEB RESOURCES

In addition to the metasites, there are many web pages devoted to particular aspects of biotechnology and genetic engineering. Here is a small sampling of the most interesting and extensive ones:

- The Biotech Chronicles, http://www.gene.com/AE/AB/BC, offers short, readable online documents on aspects of the history of biotechnology, biographical sketches of figures in the history of biotechnology and genetics, explanations of key concepts and methods in the field, a chronology, and more.
- Classic Papers in Genetics, http://www.esp.org, offers original manuscripts of some of the most important papers in the history of genetics research. It also has a chronology and links to related sites.
- Biofind, http://www2.biofind.com/home/, claims to be the Internet's "one-stop shop" for worldwide biotechnoology industry information, gossip, and news.
- The Biotechnology Information Center, http://www.nal.usda.gov/bic, is part of the National Agricultural Library, a service of the U.S. Department of Agriculture. It provides bibliographies and other publications as well as site links related primarily to agricultural biotechnology, including government regulations and patent information.
- The University of Reading in Great Britain has an excellent site at http://134.225.167.114/NCBE/GMFOOD/menu.html, offering British government and other reports and regulations related to genetically modified foods, a very controversial subject in Europe.
- The National Human Genome Research Institute, which carries out the Human Genome Project, has a variety of resources related to the project, including reports, a "talking glossary," mapping and sequence databases, a list of U.S. genome centers, and links to other genomic and genetic resources on the web. All this is accessible from http://www.nhgri.nih.gov/ For reports by the project's Ethical, Legal, and Social Implications (ELSI) Working Group, see http://www. nhgri.nih.gov/Elsi.
- The National Center for Biotechnology Information, http://130.14. 22.107/, provides access to a variety of human genome and other gene sequence databases.
- The National Library of Medicine, http://igm.nlm.nih.gov/, provides Bioethicsline, an online medical ethics database; It offers annotated bibliographies on ethics of topics such as gene therapy and human cloning.
- The National Bioethics Advisory Commission site, http://www. bioethics.gov, has not only the NBAC's own reports but links to other bioethics sites.

- The U.S. Department of Agriculture's Animal and Plant Health Inspection Service site at http://www.aphis.usda.gov/biotech/OECD/usregs.htm gives links to government agencies and regulations related to biotechnology.
- The Illinois State Academy of Science has an extensive list of links to journals and organizations in genetics at http://demeter.museum.state.il.us/isas/health/genelink.html.And here are two just for fun:

And here are two just for fun:

> Virtual FlyLab, http://vcourseware.calstatela.edu, lets you play research geneticist, mating fruit flies, analyzing the resulting offspring, and determining the laws of genetic inheritance . . . without the mess or the smell of the real thing!

> Cells Alive, www.cellsalive.com, produces pictures, including some videos, of different kinds of cells, crystals and more, with explanatory text. It enables the viewer to see heart cells beating onscreen, for example.

BIBLIOGRAPHIC RESOURCES

Bibliographic resources generally include catalogs, indexes, bibliographies, and other guides that identify the books, periodical articles, and other printed resources that deal with a particular subject. They are essential tools for the researcher.

LIBRARY CATALOGS

Most public and academic libraries have replaced their card catalogs with online catalogs, and many institutions now offer remote access to their catalog, either through dialing a phone number with terminal software or connecting via the Internet.

Access to the largest library catalog, that of the Library of Congress, is available at http://catalog.loc.gov/. This page explains the different kinds of catalogs and searching techniques available.

Yahoo! offers a categorized listing of libraries at http://dir.yahoo.com/reference/libraries/. Of course one's local public library (and for students, the high school or college library) is also a good source for help in using online catalogs.

With traditional catalogs, lack of knowledge of appropriate subject headings can make it difficult to make sure the researcher finds all relevant materials. Online catalogs, however, can be searched not only by author, title, and subject, but also by matching keywords in the title. Thus a title search for "biotechnology" will retrieve all books that have that word somewhere in their title. (Of course a book about biotechnology may not have the word *biotechnology* in the title, so it is still necessary to use subject headings to get the most comprehensive results.)

The Library of Congress catalog has numerous listings under both Biotechnology and Genetic Engineering. Clicking on "Suggestions for Related Searches" produces a list of related topics and narrower topics. The listing for Genetic Engineering, for instance, directs one to the following related topics:

- Biotechnology
- Transgenic organisms
- Genetic recombination
- Genetic intervention

It also lists the following narrower topics:

- Animal genetic engineering
- Cell nuclei—transportation
- Cloning
- Fertilization in vitro
- Gene targeting
- Gene therapy
- Microbial genetic engineering
- Molecular cloning
- Plant genetic engineering
- Protein engineering
- Recombinant DNA

Once the record for a book or other item is found, it is a good idea to see what additional subject headings and name headings have been assigned. These in turn can be used for further searching.

BOOKSTORE CATALOGS

Many people have discovered that online bookstores such as Amazon.com (http://www.amazon.com) and Barnes & Noble (http://www.barnesandnoble.com) are convenient ways to shop for books. A less-known benefit of online bookstore catalogs is that they often include publisher's information, book reviews, and reader's comments about a given title. They can thus serve as a form of annotated bibliography.

On the other hand, a visit to one's local bookstore also has its benefits. While the selection of titles available is likely to be smaller than that of an online bookstore, the ability to physically browse through books before buying them can be very useful.

PERIODICAL DATABASES

Most public libraries subscribe to database services such as InfoTrac that index articles from hundreds of general-interest periodicals (and some moderately specialized ones). The database can be searched by author or by words in the title, subject headings, and sometimes words found anywhere in the article text. Depending on the database used, "hits" in the database can result in just a bibliographical description (author, title, pages, periodical name, issue date, etc.), a description plus an abstract (a paragraph summarizing the contents of the article), or the full text of the article itself.

Many libraries provide dial-in, Internet, or telnet access to their periodical databases as an option in their catalog menu. (Telnet is a facility that allows users to run programs, such as those that access databases, from computers remote to their own.) However, licensing restrictions usually mean that only researchers who have a library card for that particular library can access the database (by typing in their name and card number). Check with local public or school libraries to see what databases are available.

A somewhat more time-consuming alternative is to find the web sites for magazines likely to cover a topic of interest. (See Specific Internet/Web Resources, p. 139, for some sites that link to journals.) Some scholarly publications are putting all or most of their articles online. Popular publications tend to offer only a limited selection. Some publications of both types offer archives of several years' back issues that can be searched by author or keyword.

FINDING ORGANIZATIONS AND PEOPLE

Lists of organizations connected with biotechnology can be found on archive sites such as Muritech and the Illinois State Academy of Sciences site (see Specific Internet/Web Resources, p. 139) and index sites such as Yahoo! If such sites do not yield the name of a specific organization, the name can be given to a search engine. Put the name of the organization in quotation marks.

Another approach is to take a guess at the organization's likely web address. For example, the American Civil Liberties Union (which includes genetic discrimination among its concerns) is commonly known by the acronym ACLU, so it is not a surprise that the organization's web site is at www.aclu.org. (Note that noncommercial organization sites normally use the .org suffix, government agencies use .gov, educational institutions have. .edu, and businesses use .com.) This technique can save time, but it doesn't always work.

There are several ways to find a person on the Internet:

Put the person's name (in quotes) in a search engine and possibly find that person's home page on the Internet.

Contact the person's employer (such as a university for an academic, or a corporation for a technical professional). Most such organizations have web pages that include a searchable faculty or employee directory.

Try one of the people-finder services such as Yahoo People Search (http://people.yahoo.com) or BigFoot (www.bigfoot.com). This may yield contact information such as an e-mail address, regular address, and/or phone number.

LEGAL RESEARCH

As issues related to biotechnology, genetic engineering, and human genetics continue to capture the attention of legislators and the public, a growing body of legislation and court cases is emerging. Because of the specialized terminology of the law, legal research can be more difficult to master than bibliographical or general research tools. Fortunately, the Internet has also come to the rescue in this area, offering a variety of ways to look up laws and court cases without having to pore through huge bound volumes in law libraries (which may not be accessible to the general public, anyway). For a start, The National Biotechnology Information Facility (http://nbif.org/) has links to just about any database or reference source for laws or court cases related to biotechnology that a researcher might need.

FINDING LAWS

When federal legislation passes, it becomes part of the United States Code, a massive legal compendium. Laws can be referred to either by their popular name or by a formal citation. For example, the DNA Identification Act is cited as 42 USC 14131, meaning title 42 of the U.S.C. code, section 14131.

The U.S. Code can be searched online in several locations, but the easiest site to use is probably the U.S. Code database at http://uscode.house.gov/. The U.S. Code may also be found at Cornell Law School (a major provider of free online legal reference material), at (http://www4.law.cornell.edu/uscode/). The fastest way to retrieve a law is by its title and section citation, but phrases and keywords can also be used.

Federal laws are generally implemented by a designated agency that writes detailed rules, which become part of the Code of Federal Regulations (C.F.R.). A regulatory citation looks like a U.S. Code citation and takes the form vol. C.F.R. sec. number, where *vol.* is the volume number and *number* is the section number. Regulations can be found at the web site for the relevant government agency, such as the Federal Trade Commission or Federal Communications Commission.

Many states also have their codes of laws online. The Internet Law Library has a page of links to state laws. The library can be accessed at a number of sites, including http://www.lectlaw.com/inll/1.htm.

KEEPING UP WITH LEGISLATIVE DEVELOPMENTS

Many bills related to such subjects as human cloning and discrimination on the basis of genetic testing are proposed in Congress and state legislatures each year. The Library of Congress catalog site (telnet locis.loc.gov) includes files summarizing legislation by the number of the Congress (each two-year session of Congress has a consecutive number: for example, the 105th Congress was in session in 1997 and 1998. Legislation can be searched for by the name of its sponsor(s), the bill number, or by topical keywords.

The Library of Congress THOMAS site (http://thomas.loc.gov/) provides a web-based interface that may be easier to use for many purposes. Summaries of legislation considered by each Congress can be searched by keyword or bill number. For example, if the researcher has read about a bill banning human cloning but doesn't know the bill number, a keyword search for "human cloning" will produce appropriate listings.

Clicking on the bill number of a particular listing gives a screen with links to summary, text of legislation, current status, floor actions, and so on. Of course, if one knows the number, one can go directly to this listing from the search screen by searching by number.

FINDING COURT DECISIONS

Like laws, legal decisions are organized using a system of citations. The general form is: *Party 1 v. Party 2 volume court reports (year)*.

Here are some examples from Chapter 2:

Bragdon v. Abbott, 97 U.S. 156 (1998)

Here the parties are Bragdon (plaintiff) and Abbott (defendant), the case is in volume 97 of the *U.S. Supreme Court Reports*, and the year the case was decided is 1998. (For the Supreme Court, the name of the court is omitted).

John Moore v. Regents of California, 51 Cal. 3d 120 (1990)

Here the parties are John Moore (plaintiff) and the Regents of the University of California (defendant), the decision is in volume 51 of the California Supreme Court records, and the case was decided in 1990.

To find a federal court decision, first ascertain the level of court involved: district (the lowest level, where trials are normally held), circuit (the main court of appeals), or the Supreme Court. The researcher can then go to a number of places on the Internet to find cases by citation and, often, the names of the parties. Two of the most useful sites include the following:

The Legal Information Institute (http://supct.law.cornell.edu/supct/) has all Supreme Court decisions since 1990 plus 610 of "the most important historic" decisions. It also links to other databases with early court decisions.

Washlaw Web (http://www.washlaw.edu/) has a variety of courts (including states) and legal topics listed, making it a good jumping-off place for many sorts of legal research.

For more information on conducting legal research, see the "Legal Research FAQ" at http://www.eff.org/pub/Legal/law_research.faq. This also explains more advanced techniques such as "Shepardizing" (referring to *Shepard's Case Citations*), which is used to find how a decision has been cited in subsequent cases and whether the decision was later overturned.

CHAPTER 7

—————————

ANNOTATED BIBLIOGRAPHY

Hundreds of books and thousands of articles and Internet documents relating to biotechnology have appeared in recent years as the field has grown in complexity and importance. They range from extremely technical "how-to" articles and descriptions of particular advances to reviews and opinion pieces aimed at the general public. This bibliography lists a representative sample of serious nonfiction sources dealing with biotechnology/genetic engineering in general, human genetics in general (for instance, discussions of the implications of the Human Genome Project, which often cover both genetic discrimination and gene therapy), and with the particular subjects within the field discussed in Chapter 1 (including a separate section on human cloning). Sources have been selected for clarity and usefulness to the general reader, recent publication (except for some items containing material of historical interest, most material dates from 1995 or later), and variety of points of view.

Listings in this bibliography are grouped according to subject and, within a subject, by type (books, articles, Internet/web documents, and other media). Newspaper and magazine articles available on the Internet are listed under Articles, not under Internet documents.

GENERAL MATERIAL ON BIOTECHNOLOGY AND GENETICS

BOOKS

Aldridge, Susan. *The Thread of Life: The Story of Genes and Genetic Engineering*. New York: Cambridge University Press, 1996. Aldridge, a biochemist, explains in simple language the world of genes and DNA, how genes can be altered, and present and future applications of genetic engineering.

Bains, William. *Biotechnology from A to Z*. New York: Oxford University Press, 1998. Describes some 1,000 biotechnology terms for nonspecialists, using clear language and concrete examples.

Barnum, Susan R., and Carol M. Barnum. *Biotechnology: An Introduction*. Belmont, Calif.: Wadsworth, 1998. Describes the nature of modern biotechnology and genetic engineering and pictures the new field's many branches, including plant and animal biotechnology, medical biotechnology and the study of the human genome, and forensic DNA profiling. Includes a discussion of regulation.

Becker, Gerhold K., and James P. Buchanan, eds. *Changing Nature's Course: The Ethical Challenge of Biotechnology*. Hong Kong: Hong Kong University Press, 1996. Collection of papers from a 1993 symposium, "Biotechnology and Ethics: Scientific Liberty and Moral Responsibility."

Bryan, Jenny. *Genetic Engineering*. Austin, Tex.: Raintree/Steck-Vaughn, 1997. Young adult book provides both a history and a current description of this rapidly advancing scientific field, including consideration of the many ethical issues it raises. Topics include genetically engineered foods, patenting of genes, DNA testing to obtain identification and health profiles, beliefs about genes' influence on behavior, and human cloning.

Bud, Robert, and Mark F. Cantley. *The Uses of Life: A History of Biotechnology*. New York: Cambridge University Press, 1993. Although genetic engineering did not become possible until the 1970s, biotechnology has been an integral part of the history of the entire twentieth century. This book describes biotechnology's 19th-century origins and key events and personalities in its history through the 1980s.

Carmen, Ira H. *Cloning and the Constitution: An Inquiry into Governmental Policymaking and Genetic Experimentation*. Madison: University of Wisconsin Press, 1985. Historically interesting book includes a review of science in American constitutional history, scientific experimentation in relation to the First Amendment, early regulation of gene cloning and experimentation, and the results of questionnaires about regulation sent to scientists and regulators.

Cole-Turner, Ronald. *The New Genesis: Theology and the Genetic Revolution*. Louisville, Ky.: Westminster John Knox, 1993. Shows how Christian theology can provide a framework for considering issues raised by biotechnology, including limits that should be set on genetic alteration of humans and other living things.

Drlica, Karl. *Understanding DNA and Gene Cloning: A Guide for the Curious*. New York: Wiley, 1996. Using terms and analogies easily understandable by nonscientists, author explains basic molecular biology, manipulation of DNA, insights gained through gene cloning, and human genetics.

Frank-Kamenetskii, Maxim D., and Lev Liapin. *Unraveling DNA: The Most*

Important Molecule of Life. Reading, Mass.: Perseus, 1997. Describes the history of basic research on genetics and DNA, how DNA works in cells, and recent advances, including attempts at gene therapy for cancer and AIDS.

Gonick, Larry, and Mark Wheelis. *The Cartoon Guide to Genetics*. New York: Harper, 1991. An enjoyable introduction to the principles and major discoveries in genetics that underlie biotechnology.

Gottweis, Herbert. *Governing Molecules: The Discursive Politics of Genetic Engineering in Europe and the United States*. Cambridge, Mass.: MIT Press, 1998. Complex political history explains how genetic engineering became such a controversial technology, arousing extremes of positive and negative emotion, and provides evidence that many of the promises of biotechnology may be mere wishful thinking, whereas causes of concern about it seem to be real. Covers the United States but emphasizes events in Europe.

Grace, Eric S. *Biotechnology Unzipped: Promises and Realities*. Washington, D.C.: Joseph Henry Press, 1997. Clearly written chapters cover how biotechnology came about, tools in the genetic engineering workshop, medical biotechnology, agricultural biotechnology, biotechnology and the environment, biotechnology in the natural world, and ethical issues.

Grobstein, Clifford. *A Double Image of the Double Helix: The Recombinant-DNA Debate*. San Francisco: W. H. Freeman, 1979. Describes the first round of genetic engineering and attempts to regulate it in the early 1970s; reprints several important sets of guidelines.

Ho, Mae-Wan. *Genetic Engineering: Dream or Nightmare?* Bath, Britain: Gateway Books, 1998. Author considers genetic engineering to be an untried and inadequately researched technology that has gotten out of control, a mixture of "bad science and big business" that must be stopped.

Holdrege, Craig. *Genetics and the Manipulation of Life: The Forgotten Factor of Context*. Hudson, N.Y.: Lindisfarne Books, 1996. Puts recent discoveries in biotechnology and genetic engineering into a broad scientific, social, and ethical context to help the lay reader make educated judgments about the issues these new technologies raise.

Judson, Horace Freeland. *The Eighth Day of Creation: The Makers of the Revolution in Biotechnology*. New York: Simon & Schuster, 1979. Detailed, interesting account of Watson and Crick's determination of the structure of the DNA molecule and other key midcentury discoveries in genetics.

Kornberg, Arthur. *The Golden Helix: Inside Biotech Ventures*. Mill Valley, Calif.: University Science Books, 1996. A Nobel Prize–winning scientist whose work with enzymes and DNA helped to make modern genetic engineering possible offers an insider's view of biotechnology as both science and industry, including portraits of several companies and their executives.

Annotated Bibliography

Lagerkvist, Ulf. *DNA Pioneers and Their Legacy*. New Haven: Yale University Press, 1988. Describes Watson, Crick, and others whose discoveries set the stage for modern molecular biology and genetic engineering.

Levine, Joseph, and David Suzuki. *The Secret of Life: Redesigning the Living World*. New York: W. H. Freeman, 1998. Companion volume to an eight-part PBS television series first presented in 1993 explains advances in molecular biology, genetic engineering, and gene therapy and discusses how they are changing medicine and society.

Marsa, Linda. *Prescription for Profits: How the Pharmaceutical Industry Bankrolled the Unholy Marriage Between Science and Business*. New York: Scribner, 1997. Investigative reporter provides evidence to show that the research process is often corrupted when researchers ally themelves with drug and biotechnology companies.

Meyers, Robert A., ed. *Molecular Biology and Biotechnology: A Comprehensive Desk Reference*. New York: Wiley, 1995. Brief but thorough entries should be able to answer just about any question relating to molecular biology, biotechnology, or molecular medicine. Contains more than 250 articles by outstanding scientists.

Miller, Henry I. *Policy Controversy in Biotechnology: An Insider's View*. Georgetown, Tex.: R. G. Landes Co., 1997. Miller, a senior research fellow at the Hoover Institution, draws on his experience as a Food and Drug Administration official from 1979 to 1994. He feels that regulation of the biotechnology industry has been excessive.

O'Mahoney, Patrick, ed. *Nature, Risk and Responsibility: Discourses on Biotechnology*. New York: Routledge, 1999. Collection of essays draws on a wide range of social theories to explore the ethical issues raised by biotechnology and its implications for nature, life, and social organization. Contributors evaluate risks and propose key strategies for the future.

Peters, Ted. *Playing God? Genetic Determinism and Human Freedom*. New York: Routledge, 1997. Discusses the theological dilemmas presented by recent advances in genetic engineering, including the patenting of genetic information.

Rabinow, Paul. *Making PCR: A Story of Biotechnology*. Chicago: University of Chicago Press, 1997. Rabinow, an anthropologist, studied the scientists working at Cetus, a major biotechnology company. He describes both the benefits and the conflicts over prestige and money that occurred in the wake of the company's development of PCR (polymerase chain reaction), a groundbreaking technique for reproducing small amounts of DNA.

Raleff, Tamara L., ed. *Biomedical Ethics: Opposing Viewpoints*. San Diego, Calif.: Greenhaven Press, 1998. Pro and con essays on biological patents, ownership of body cells/tissues, reproductive technologies, human genetic research and gene modification, and uses and risks of human genetic testing.

Biotechnology and Genetic Engineering

Reiss, Michael J., and Roger Straughan. *Improving Nature: The Science and Ethics of Genetic Engineering*. New York: Cambridge University Press, 1999. A biologist and a philosopher provide a balanced description of the science of genetic engineering in microorganisms, plants, animals, and humans and consideration of the ethical questions it raises. The book includes case studies and recommendations for education.

Rifkin, Jeremy. *The Biotech Century: Harnessing the Gene and Remaking the World*. New York: Putnam, 1998. Rifkin, widely known for his negative views of biotechnology and genetic engineering, warns of the dangers of such things as the patenting of genetically engineered life forms, human gene alteration, and possible human cloning and urges restraint, responsibility, and caution in the use of new genetic technologies.

Rudolph, Frederick B., and Larry V. McIntire, eds. *Biotechnology: Science, Engineering, and Ethical Challenges for the 21st Century*. Washington, D.C.: Joseph Henry Press, 1996. Collection of essays covering the research that founded modern biotechnology, present and near-future biotechnology applications in various fields (mostly medical), technology transfer, ethics, and the role of government in the development of biotechnology.

Smith, John E. *Biotechnology* (3rd ed.). New York: Cambridge University Press, 1996. Describes a wide range of biotechnology products, including fuels, fermentation, enzymes, single-cell proteins, medical products, environmental cleanup and enrichment products, agriculture and forestry products, and foods and beverages; also considers techniques, patenting, safety, and ethical issues.

Thackray, Arnold. *Private Science: Biotechnology and the Rise of the Molecular Sciences*. Philadelphia: University of Pennsylvania Press, 1998. Collection of essays focuses on the relationships among corporations, universities, and national governments involved in biochemical research; covers historical, political, and economic aspects.

Turney, Jon. *Frankenstein's Footsteps: Science, Genetics, and Popular Culture*. New Haven: Yale University Press, 1998. Describes how images from popular novels such as *Frankenstein* and *Brave New World* have both reflected and shaped popular distrust of biotechnology. Calls for new stories that take a more intelligent and positive approach.

Vandermeer, John. *Reconstructing Biology: Genetics and Ecology in the New World Order*. New York: Wiley, 1996. Claims that the "trinity of genetic determinism, neomalthusianism, and nature worship" had been socially constructed by popular culture and biologists; considers the implications for views of the environment, race, evolution, genetics, gender roles, and other topics.

Van Dijck, José. *Imagenation: Popular Images of Genetics*. New York: New York University Press, 1998. Describes the exaggerated positive and negative pictures that dominate popular images of biotechnology/genetic engineering and the present and future effect of these technologies on society; suggests how to "retool the imagination" for a more accurate view.

Wade, Nicholas, ed. *The Science Times Book of Genetics*. New York: Lyons Press, 1999. Fascinating collection of articles from *The New York Times* science section on recent advances in genetics and genetic engineering.

Watson, James D. *The Double Helix*. New York: Atheneum, 1968. Lively, if biased, memoir of the landmark discovery of the structure of DNA by one of the codiscoverers.

———, and John Tooze. *The DNA Story: A Documentary History of Gene Cloning*. San Francisco: W. H. Freeman, 1981. Useful collection of original material (including contemporary magazine articles) on controversy over and regulation of the first round of genetic engineering.

Wekesser, Carol, ed. *Genetic Engineering: Opposing Viewpoints*. San Diego: Greenhaven, 1996. Anthology of essays pro and con on whether genetic engineering benefits or harms society, how genetic engineering affects agriculture, whether DNA identification is accurate, how genetic engineering may affect health care, and how genetic engineering should be regulated.

Wheale, Peter R., and Ruth M. McNally. *Genetic Engineering: Catastrophe or Utopia?* New York: St. Martin's Press, 1988. Includes much specific detail on regulatory policy, patenting, release of genetically engineered organisms, human genetic screening, and human gene modification/eugenics through the late 1980s.

Yount, Lisa. *Genetics and Genetic Engineering*. New York: Facts On File, 1997. For young adults. Profiles key discoveries in genetics and genetic engineering and the people who made them; also considers their social implications.

ARTICLES

Beardsley, Tim. "Smart Genes." *Scientific American*, vol. 265, August 1991, pp. 86–95. Describes genes, organisms, and environment as components of a complex network that constantly influence each other; argues against the notion that information and control move only from genes to organisms.

Carey, John, and Julia Flynn. "The Biotech Century." *Business Week*, March 10, 1997, pp. 78ff. The deciphering of the genomes of humans and other living things is likely to make the 21st century the "biotech century" and transform not only medicine but agriculture, computers, and many other industries. It will also present major legal and moral challenges.

Enriquez, Juan. "Genomics and the World's Economy." *Science*, vol. 281, August 14, 1998, pp. 925–926. Genomics allows scientists to study, design, and build biologically important molecules. It is restructuring major companies, forcing them to become more interdisciplinary, and changing the world economy.

"Evolution on Fast-Forward." *Business Week*, no. 3648, September 27, 1999, p. 140. Through the process of directed evolution, or breeding at the molecular level, scientists can greatly improve qualities of genetically created compounds within weeks and even develop totally new substances.

Gibbons, Ann. "Biotech's Second Generation." *Science*, vol. 256, May 8, 1992, pp. 766–768. Biotechnologists in the early 1990s were beginning to design drugs that improved on nature rather than just mimicking natural compounds. Article gives a good overview of biotechnology advances in this period.

Gurin, Joel, and Nancy E. Pfund. "Bonanza in the Bio Lab." *The Nation*, vol. 231, November 22, 1980, pp. 543–548. Describes biotechnology's first spectacular commercial successes and some of the safety and patenting/intellectual property issues they raised.

Holz, Robert Lee. "Caught in a Furor of His Own Creation." *Los Angeles Times*, March 6, 1997, pp. A1, A20. Portrait of Scottish scientist Ian Wilmut, who found himself famous after announcing that he had cloned a sheep from a mature cell taken from an adult animal.

Huff, Toby E. "The Fourth Scientific Revolution." *Society*, vol. 33, May–June 1996, pp. 9ff. Considers environmental, religious, ethical, and legal consequences of genetic engineering, including animal and human genetic engineering, creation of new organisms, and human gene therapy.

Keeler, Robert. "Uses for PCR Are Multiplying in Gene-Related Research." *R&D*, vol. 33, August 1991, p. 30ff. Describes the polymerase chain reaction (PCR), a process that allows tiny amounts of DNA to be multiplied quickly to make enough material for testing, and its many uses, including gene sequencing in the Human Genome Project and analysis of DNA found at crime scenes.

Lutz, Diana. "Hello, Hello, Dolly, Dolly." *The Sciences*, May/June 1997, pp. 10–11. Describes a press conference held at the New York Academy of Sciences soon after the famed cloning of a sheep in 1997, in which a panel of eminent scientists pointed out that the real importance of the experiment was its demonstration that developed cells can return to their embryonic state; this discovery has major implications for biotechnology and medicine.

Miller, Julie Ann. "Lessons from Asilomar." *Science News*, vol. 127, February 23, 1985, pp. 122–123, 126. Ten years after the historic conference that produced the first guidelines for genetic engineering research, scientists looked back with a mixture of pride and disappointment.

Rifkin, Jeremy. "God in a Labcoat: Can We Control the Biotech Revolution Before It Controls Us?" *Utne Reader*, no. 87, May–June 1998, pp. 66ff. The biotechnology revolution should be approached cautiously because of its potential to alter social structures and norms, author says; the public should take steps to ensure that its effects do minimum harm.

Shreeve, James. "Secrets of the Gene." *National Geographic*, vol. 196, October 1999, pp. 42–75. Surveys recent advances in genetic analysis, primarily in humans, including study of individual genomes and family patterns.

Studt, Tim. "Gene Chip Technologies Transform Biological Research." *R&D*, vol. 40, February 1998, pp. 38ff. Improvements in techniques for making so-called gene chips (computer chips that contain and analyze DNA) have greatly increased the chips' production. The article gives a good explanation of how the chips work, how they are made, and their great economic potential.

Travis, John. "Chips Ahoy." *Science News*, vol. 151, March 8, 1997, pp.144–145. Microchips covered with DNA are emerging as powerful research tools with uses ranging from basic biological research to studies of human genetic diversity, diagnosis of disease, and identification of potentially dangerous mutated genes.

INTERNET/WEB DOCUMENTS

Bohlin, Ray. "The World View of *Jurassic Park*." Leadership University/Probe Ministries. Available online. URL: http://www.leaderu. com/orgs/probe/docs/jurassic.html. Posted in 1995. Bohlin, director of communications at Probe Ministries, uses the famous movie as a starting point from which to discuss dangers of science, biotechnology, computers, evolutionary assumptions, and the New Age.

Kohler, Gus A. *Bioindustry: A Description of California's Bioindustry and Summary of the Public Issues Affecting Its Development*. Available online. URL: http://www.library.ca.gov/CRB/96/07/index.html. Posted in April 1996. Extensive survey of biotechnology, ranging from transgenic plants and animals to federal regulations and ethical issues (safety, patenting living things, human genetic testing, etc.).

MIT Experimental Study Group. *Recombinant DNA*. Available online. URL: http://esg-www.mit.edu:8001/esgbio/rdna/rdnadir.html. Downloaded July 1, 1999. Biology hypertextbook that covers gene cloning techniques, polymerase chain reaction, and forensic DNA fingerprinting.

OTHER MEDIA

Biotechnology Today: Who Owns Genes. Video, 60 minutes. Bethesda, Md.: Mentor Media/Bioconferences International, Inc., 1993. Two half-hour PBS specials, the first exploring recent developments in biotechnology through interviews and the second consisting of a panel of experts discussing agricultural technology transfer and biotechnology patent issues.

AGRICULTURAL BIOTECHNOLOGY AND SAFETY

Note: Articles about Dolly, the cloned sheep whose existence was announced in 1997, appear here if they deal primarily with that advance in itself or its implications for animal cloning. Articles that deal primarily with Dolly's implications for human cloning are in the Human Cloning section of this bibliography.

BOOKS

Dawkins, Kristin. *Gene Wars: The Politics of Biotechnology.* New York: Seven Stories Press, 1997. Short book expresses fears that multinational corporations will control the genetic engineering of food plants, dangerously reducing genetic diversity and allowing millions to go hungry.

DeSalle, Rob, and David Lindley. *The Science of Jurassic Park and the Lost World: Or, How to Build a Dinosaur.* New York: Harper, 1997. Considers whether it would be possible to recreate dinosaurs or other extinct animals from bits of their DNA, as envisioned in the popular movies.

Doyle, J. C., and G. J. Persley, eds. *Enabling the Safe Use of Biotechnology: Principles and Practice.* Washington, D.C.: World Bank, 1996. Provides a practical guide for policymakers and others attempting to make decisions about the safe use of modern biotechnology. Concludes that risks from biotechnology are low.

Durant, J., M. W. Bauer, and G. Gaskell, eds. *Biotechnology in the Public Sphere: A European Sourcebook.* San Luis Obispo, Calif: Cromwell Press, 1998. Provides information on the public controversy about biotechnology and genetically engineered foods in Europe.

Fowler, Cary, and Pat Mooney. *Shattering: Food, Politics, and the Loss of Genetic Diversity.* Tucson: University of Arizona Press, 1990. Warns that genetically engineered crops are increasing the loss of plant genetic diversity begun by the world emphasis on monoculture; recommends saving "heirloom" seeds to preserve diversity.

Annotated Bibliography

Genetically Modified Foods: Benefits and Risks, Regulation and Public Acceptance. London: Parliamentary Office of Science and Technology, 1998. British government report responding to public concern about genetically engineered foods.

Hallberg, M., ed. *Bovine Somatotropin and Emerging Issues.* Boulder, Colo.: Westview Press, 1992. Describes some of the undesirable side effects of administering genetically engineered bovine growth hormone to dairy cattle and offers procedures for anticipating and evaluating similar problems with other bioengineered substances.

Hambleton, P., J. Melling, and T. Salisbury, eds. *Biosafety in Industrial Biotechnology.* London and New York: Blackie Academic and Professional, 1994. Describes local, national, and international regulations, agencies, and policymaking bodies that affect industrial biotechnology and offers a practical approach to dealing with them. Includes both industrial safety and environmental monitoring.

Institute of Food Technologies. *Appropriate Oversight for Plants with Inherited Traits for Resistance to Pests.* Chicago: Institute of Food Technologies, 1996. Recommends principles for regulating genetically altered plants. Argues that current EPA regulations are too strict and should focus on inherent risks of the organisms, not on the fact that they are genetically engineered.

Juma, Celestous. *The Gene Hunters: Biotechnology and the Scramble for Seeds.* Princeton, N.J.: Princeton University Press, 1989. Describes biotechnology companies' hunt for wild varieties of plants with the potential to improve farm crops.

Kneen, Brewster. *Farmageddon: Food and the Culture of Biotechnology.* New Society Publications, 1999. Claims that multinational conglomerates are genetically engineering food plants to make a profit, regardless of possible risks to human health or the environment.

Kolata, Gina. *Clone: The Road to Dolly and the Path Ahead.* New York: Morrow, 1998. Describes animal cloning research up to and including Dolly the sheep, providing a good feel for the process of science and the life of a laboratory scientist. Last chapter considers ethical issues raised by animal and human cloning.

Krimsky, Sheldon, and Roger P. Wrubel. *Agricultural Biotechnology and the Environment: Science, Policy, and Social Issues.* Champaign: University of Illinois Press, 1996. Describes hotly debated advances in agricultural biotechnology, such as giving genetically engineered growth hormone to cattle and adding genes to plants to make them able to produce a bacterial pesticide or to resist chemical herbicides.

Lappe, Marc, and Britt Bailey. *Against the Grain: Biotechnology and the Corporate Takeover of Your Food.* Monroe, Maine: Common Courage Press,

1999. Maintains that, contrary to the claims of biotechnology companies, genetically modified foods may present dangers to human health.

Marshall, Elizabeth L. *High-Tech Harvest: A Look at Genetically Engineered Foods*. Danbury, Conn.: Franklin Watts, 1999. Young-adult book looks at present and probable future advances in genetically engineered foods and the ethical, health, and environmental issues they raise.

Mather, Robin. *A Garden of Unearthly Delights: Bioengineering and the Future of Food*. New York: Plume, 1995. Paints a mostly negative picture of a future filled with genetically manipulated livestock and produce. Includes the viewpoints of farmers and researchers.

Miller, Henry I. *Policy Controversy in Biotechnology*. San Diego, Calif.: Academic Press, 1997. Critiques regulation of biotechnology in the United States by federal agencies including the FDA, EPA, USDA, and NIH; claims that "government is the problem" and offers suggestions for reform.

Nottingham, Stephen. *Eat Your Genes: How Genetically Modified Food Is Entering Our Diet*. New York: St. Martin's Press, 1998. Considers both the intended benefits and the unintended side effects of genetically engineered foods. Also discusses ethical issues, such as humane treatment of transgenic animals raised for food or "pharming."

Oei, Hong Lim. *Genes and Politics: The Recombinant DNA Debate*. Burke, Va: Chatelaine Press, 1998. Describes technocratic, democratic, and legislative policymaking processes in the early years of recombinant DNA research and describes lessons that debates on the subject in the 1970s have for biotechnology policy today; makes recommendations for a better policymaking process.

Rissler, Jane, and Margaret Mellon. *The Ecological Risks of Engineered Crops*. Cambridge, Mass.: MIT Press, 1996. Two members of the Union of Concerned Scientists explain the organization's concerns about risks presented by transgenic crops and offer recommendations.

Rollin, Bernard E. *The Frankenstein Syndrome: Ethical and Social Issues in the Genetic Engineering of Animals*. New York: Cambridge University Press, 1995. A balanced and nontechnical discussion of ethical issues raised by the genetic engineering of animals, describing both benefits and dangers. Author recommends strong regulation of the field.

Sayler, Gary S., John Sanseverino, and Kimberly L. Davis, eds. *Biotechnology in the Sustainable Environment*. New York: Plenum Press, 1997. Focuses on techniques for using biotechnology to clean up the environment (bioremediation). Also describes ways that biotechnology can reduce environmental damage by substituting for existing technologies.

Select Committee on the European Communities of the House of Lords. *EC Regulation of Genetic Modification in Agriculture*. London: H. M. Stationery

Office, 1998. Describes current European Union regulation of genetically modified crops and foods and makes suggestions for modifications.

Serageldin, Ismail, and Wanda Williams Collins, eds. *Biotechnology and Biosafety*. Washington, D.C.: World Bank, 1998. Proceedings of a meeting during the fifth annual World Bank Conference on Environmentally and Socially Sustainable Biotechnology.

Shapiro, Robert B. *Trade, Feeding the World's People, and Sustainability: A Cause for Concern*. St. Louis, Mo.: Center for the Study of American Business, Washington University, 1998 (CEO series no. 22). Shapiro, CEO of Monsanto Corporation, maintains that biotechnology offers the best hope of feeding the world's growing population without seriously damaging the environment.

Shiva, Vandana. *Biopiracy: The Plunder of Nature and Knowledge*. Boston: South End Press, 1997. Argues for a "collective intellectual property right" that would give the people who live in a region all profits from use of a natural resource that comes from that region, even if the resource is made into a useful product in another place.

Ticciati, Laura, and Robin Ticciati. *Genetically Engineered Foods: Are They Safe? You Decide*. New Canaan, Conn.: Keats, 1998. A short book describing possible dangers of genetically engineered foods, which growing numbers of consumers are eating unknowingly because such foods do not have to be labeled in the United States.

U.S. Government. *Use of Bovine Somatotropin (BST) in the United States: Its Potential Effects*. Washington, D.C.: Government Printing Office, 1994. Study conducted by the Executive Branch evaluates the use of genetically engineered bovine growth hormone to increase milk production in dairy cattle; does not consider socioeconomic effects.

ARTICLES

Abelson, Philip H. "The Third Technological Revolution." *Science*, vol. 279, March 27, 1998, p. 2019. Although the medical side of the new science of genomics, based on detailed analysis and changing of genes, has so far been emphasized, the greatest global impact of genomics is likely to come from manipulation of plant genes. Genetically engineered plants will probably provide most of the world's food, fuel, and fiber in the future.

"The Age of Dolly." *Newsweek*, June 7, 1999, p. 70. By at least one standard (the number of bodies called telomeres lost from the tips of her cells' chromosomes), Dolly the cloned sheep appears to be as old as her donor "mother," six years older than her own birth age. This news dims future prospects for both animal and human clones.

Biotechnology and Genetic Engineering

Anderson, Gary B., and George E. Seidel. "Cloning for Profit." *Science*, vol. 280, May 29, 1998, pp. 1400–1401. Although media accounts of the cloning of a sheep focused mainly on ethical issues raised by possible cloning of humans, the chief near-term application of the technique will be to efficiently generate transgenic farm animals containing specific, commercially useful genes, as is already done routinely with plants.

Baker, Beth. "Streamlining of Biotech Regulations Pleases Industry." *BioScience*, vol. 44, September 1994, p. 527. The biotechnology industry has expressed approval of the U.S. government's decision that field testing of genetically engineered crops will no longer require government supervision, although the Department of Agriculture must still be notified of such tests.

Baskin, Yvonne. "Getting the Bugs Out." *Atlantic Monthly*, vol. 265, June 1990, pp. 40–47. Describes the first release of genetically engineered organisms into the environment: Steven Lindow's release of "ice-minus" bacteria, intended to make plants frost resistant, in California in 1987.

———. "Into the Wild." *Natural History*, vol. 108, October 1999, pp. 34–37. Studies showing that bioengineered squash can pass inserted genes to nearby wild relatives increases fears that genetically engineered crops may do serious ecological damage.

Begley, Sharon. "Little Lamb, Who Made Thee?" *Newsweek*, vol. 129, March 10, 1997, pp. 52–60. The cloning of a sheep from the body cell of an adult ewe in Scotland is a breakthrough in genetic engineering, but it raises ethical issues, especially concern about possible human cloning.

———. "Spring Cloning." *Newsweek*, vol. 129, June 30, 1997, pp. 82–83. Three months after the cloning of Dolly the lamb made headlines, other scientists announced that pigs, cows, and other types of livestock had also been made pregnant with clones.

Bellow, Daniel. "Vermont, the Pure-Food State." *The Nation*, vol. 268, March 8, 1999, p. 18. Monsanto Corporation and the state of Vermont have been fighting over the use of bovine growth hormone to increase milk production in cows. The federal government considers the hormone safe, but some research has linked it with human breast and prostate cancer.

Benson, Susan, Mark Arax, and Rachel Burstein. "A Growing Concern." *Mother Jones*, vol. 22, January–February 1997, pp. 36ff. Expresses fears that genetically altered food crops are being introduced at a rate too rapid for proper monitoring and regulation; claims that engineered seeds will control some pests but create new problems with others.

Burstein, Rachel. "Paid Protection." *Mother Jones*, vol. 22, January–February 1997, p. 42. Maintains that, far from objecting to government regulation, large biotechnology companies favor it because it helps them eliminate less powerful rivals and head off criticisms from environmentalists.

Annotated Bibliography

Caldwell, Mark. "Cumulina and Her Sisters." *Discover*, vol. 20, January 1999, p. 56. The commercial viability of cloning mammals from adult cells was demonstrated when University of Hawaii scientists produced more than 60 cloned mice, including clones of clones.

Carey, John. "Barnyard Biotech Breeds High Hopes." *Business Week*, no. 3587, July 20, 1998, pp. 56–57. Dolly the sheep is no longer one of a kind; both sheep and cattle are beginning to be cloned on a greater scale, and many in the biotechnology industry believe that cloning of drug-producing animals could be very profitable.

Carruthers, Fiona. "Cooking with Genes." *Time International*, vol. 154, August 16, 1999, pp. 44–45. Until recently, Australia and New Zealand had seemed free of the fears of genetically modified food so prevalent in Europe, but now the public in these countries is beginning to debate whether genetically altered crops should be grown there or genetically modified foods labeled..

Carter, Jimmy. "Biotechnology Can Defeat Famine." *New Perspectives Quarterly*, vol. 14, November 1997, pp. 32–33. Studies show that biotechnology, in the form of new planting and weed control procedures and new seed varieties, can significantly increase harvests and thus food supplies for the world's hungry people.

Cohen, Jon. "Can Cloning Help Save Beleaguered Species?" *Science*, vol. 276, May 30, 1997, pp. 1329–1330. Zoologists are cautiously optimistic that cloning could be used to preserve rare animals in zoos, help to save species that do not breed well in captivity, and improve genetic diversity.

Couzin, Jennifer. "What's Killing Clones?" *U.S. News & World Report*, vol. 126, May 24, 1999, p. 65. Early deaths of some cloned farm animals have led researchers to realize that many technical problems remain to be solved before cloning is safe enough to try on humans, or even on animals on a large scale.

Drucker, Steven M., and L. Val Giddings. "Should All Genetically Engineered Foods Be Labeled?" *CQ Researcher*, vol. 8, September 4, 1998, p. 777. Drucker, executive director of the Alliance for Bio-Integrity, maintains that genetically engineered foods should be labeled because both law and ethics demand it, but Giddings, vice president for food and agriculture of the Biotechnology Industry Organization, says that labeling foods from engineered crops that are nutritionally indistinguishable from natural ones may mislead consumers.

Ehrenfeld, David. "A Techno-pox upon the Land." *Harper's Magazine*, vol. 295, October 1997, pp. 13ff. New biotechnology advances, such as genetically engineered plants that resist particular herbicides or produce their own insecticides, push farmers to become dependent on the large companies that produce the plants and the chemicals they need and may harm consumer health as well.

Falk, Bryce W., and George Bruening. "Will Transgenic Crops Generate New Viruses and New Diseases?" *Science*, vol. 263, March 11, 1994, pp. 1395–1396. Maintains that there is little possibility that genetically altered plants will develop new plant viruses or diseases.

"Fields of Genes." *Business Week*, no. 3624, April 12, 1999, p. 62. Maintains that, although seeds for genetically engineered crops cost more than those for unaltered ones, they are worth the price because they produce higher yields, can grow in a wide variety of environmental conditions, and can repel pests without requiring chemical pesticides.

Firth, Peta. "Leaving a Bad Taste." *Scientific American*, vol. 280, May 1999, pp. 34–35. Arpad Pusztai of the Rowett Research Institute in Aberdeen, Scotland, revealed in a hearing before the British Parliament that rats fed genetically modified potatoes suffered damaged immune systems and stunting of vital organs. His subsequent dismissal from his post has added to public outcry against genetically modified foods.

Giampietro, Mario. "Sustainability and Technological Development in Agriculture: A Critical Appraisal of Genetic Engineering." *Bioscience*, vol. 44, November 1994, pp. 677ff. Considers whether genetic engineering can be successful in allowing agriculture to sustain the world's rapidly growing population.

Glausiusz, Josie. "The Great Gene Escape." *Discover*, vol. 19, May 1998, pp. 90–96. Describes the possibility that genes, inserted into food plants, such as genes for disease control or herbicide resistance, could be transferred to weeds growing nearby; speculates on the results of such transfer.

Gordon, Meg. "Suffering of the Lambs." *New Scientist*, April 26, 1997, pp. 16–17. Claims that if cloning and production of transgenic animals become widespread, they will increase the suffering of animals used in biotechnology and medical experimentation.

Gotsch, N., and P. Rieder. "Biodiversity, Biotechnology, and Institutions Among Crops: Situation and Outlook." *Journal of Sustainable Agriculture*, vol. 5, January–February 1995, pp. 5–40. Considers the effect of agricultural genetic engineering and the holding of biotechnology patents on the biological diversity of farm crops.

Graaf, F. K. D. "Biotechnology and Sustainable Development in the Third World." *Trends in Biotechnology*, vol. 9, September 1991, pp. 297–299. Considers whether biotechnology will help small farmers and increase the chances of developing sustainable agriculture or whether it will simply put such farmers in bondage to the multinational corporations that own the patents on genetically engineered crops.

Annotated Bibliography

Green, Emily. "The Spud America Didn't Like." *New Statesman*, vol. 129, February 26, 1999, p. 18. Expresses concern over genetically engineered products such as Monsanto's New Leaf potato, which contains a bacterial gene that allows the plant to produce its own insecticide.

"Greens v. Genes." *Economist*, vol. 344, July 19, 1997, pp. 18–19. Editorial maintains that Europe's Green parties have restricted European biotechnology companies for too long because of concerns about safety and ethical issues. Editors ask that restrictions be lifted so that Europe's biotechnology industry can contribute to society as the industry has done in the United States.

Hager, Mary, and Adam Rogers. "A Biotech Roadblock." *Newsweek*, vol. 131, April 13, 1998, p. 66. With the support of anti–biotechnology activist Jeremy Rifkin, cell biologist Stuart Newman of New York Medical College is trying to obtain a patent for the making of human-animal chimeras (creatures that contain genes from more than one species) so that he can prevent others from making such creatures.

Hall, Stephen S. "One Potato Patch That Is Making Genetic History." *Smithsonian*, vol. 18, August 1987, pp. 125–136. Describes the science behind and the debate surrounding the first release of genetically engineered organisms (bacteria intended to make plants frost resistant) into the environment in 1987.

Harris, Mark. "Fresh from the Lab." *Vegetarian Times*, August 1999, pp. 58ff. Warns of allergies, ecological damage, and other potential side effects of genetically engineered food crops that may outweigh their benefits. Calls for labeling and closer regulation of genetically modified foods.

Hawaleshka, Danylo. "Unnatural Selection: Are Genetically Altered Foods Really Safe?" *Maclean's*, vol. 110, January 20, 1997, pp. 56–57. Health Canada, the Canadian equivalent of the FDA, has approved 19 genetically altered foods so far, but debate over their safety to human health and the environment continues.

Hess, C. E. "Biotechnology-Derived Foods from Animals." *Critical Review of Food Science and Nutrition*, vol. 32, February 1992, pp. 147–150. Describes why biotechnology works better than traditional animal breeding in helping American agriculture produce food that is safe, nutritious, relatively inexpensive, and environmentally sound; warns that progress may be held up by public fears of genetic engineering and by arguments over issues such as the patenting of animals.

Kahn, Jennifer. "The Green Machine." *Harper's Magazine*, vol. 298, April 1999, p. 70. Critics say that forbidding farmers to save and reuse genetically engineered seeds, as Monsanto does, gives the giant companies that market the seeds an unfair stranglehold on farmers and could threaten the world's ability to feed itself.

Kimbrell, Andrew. "Facing the Future: Genetic Engineering." *The Animals' Agenda*, vol. 15, January–February 1995, pp. 24ff. Animals containing genes from humans or other species may be useful for laboratory researchers or agribusiness, but such animals often have debilitating physiological traits that limit their survival.

Kling, James. "Could Transgenic Supercrops One Day Breed Superweeds?" *Science*, vol. 274, October 11, 1996, pp. 180–181. Describes some ecologists' fear that crops genetically engineered to resist herbicides or disease could spread their genes to nearby weeds, especially in developing countries, where crops may be grown near closely related wild plants.

Kluger, Jeffrey. "The Suicide Seeds." *Time*, vol. 153, February 1, 1999, p. 44. A grassroots protest broke out on the Internet against Monsanto Corp.'s new patented seeds, which became unable to reproduce as soon as they matured, forcing farmers to buy new seeds from the company repeatedly. (Monsanto later withdrew these "Terminator" seeds.)

Kock, Kathy. "Food Safety Battle: Organic vs. Biotech." *CQ Researcher*, vol. 8, September 4, 1998, pp. 763ff. Biotechnology companies may be endangering the survival of organic foods by buying seed companies and reengineering traditional organic pesticides and by developing genetically enhanced products that compete with organic ones. Labeling of both kinds of foods is also an issue.

Kolata, Gina. "With Cloning of a Sheep, the Ethical Ground Shifts." *New York Times*, February 24, 1997, pp. A1, C17. Describes the science behind the cloning of a lamb, Dolly, from an udder cell of an adult ewe; also discusses the ethical questions this advance raises, including those related to human cloning. Sidebar articles discuss use of the cloning technique in animal biotechnology and describe Ian Wilmut, the Scottish scientist who cloned Dolly.

"Labeling the Mutant Tomato." *Economist*, vol. 344, August 9, 1997, p. 54. The European Community now requires labeling of foods containing genetically engineered plants such as soybeans and corn.

Lenzner, Robert, and Bruce Upbin. "Monsanto v. Malthus." *Forbes*, vol. 159, March 10, 1997, pp. 58ff. Monsanto CEO Robert Shapiro has been changing his giant company's emphasis from chemicals to biotechnology and claims that this science will fuel the increase in food production needed to support the world's growing population.

Majzoub, Joseph A., and Louis J. Muglia. "Knockout Mice." *New England Journal of Medicine*, vol. 334, April 4, 1996, pp. 904ff. Describes technique for deleting both maternal and paternal copies of particular genes in mice; this technique is very useful in determining the functions of genes.

Mann, Charles. "Biotech Goes Wild." *Technology Review*, vol. 102, July–August 1999, pp. 36ff. Detailed evaluation of dangers and benefits of present and near-future genetically engineered crop plants. Calls for better-organized regulation.

Middendorf, Gerad, et al. "New Agricultural Biotechnologies: The Struggle for Democratic Choice." *Monthly Review*, vol. 50, July–August 1998, pp. 85–86. The promise of agricultural biotechnology is marred by lack of public input into the field.

Miller, Henry I. "Where Are the Promised Wonders of Biotech?" *Consumers' Research*, vol. 81, March 1998, pp. 19ff. Maintains that obstructive federal regulations, especially those on genetically engineered crops, have kept American consumers from enjoying the economic benefits of biotechnology.

Miller, Robert V. "Bacterial Gene Swapping in Nature." *Scientific American*, vol. 278, January 1998, pp. 66ff. Bacteria in nature trade genes more often than scientists had thought. Studies of this process, called conjugation, should help in predicting and preventing risks from releasing genetically engineered bacteria.

Milius, Susan. "Future Farmers May Collect Urine, Not Milk." *Science News*, vol. 153, January 10, 1998, p. 21. Scientists have created transgenic mice that produce human growth hormone in their bladders. This process could be developed to produce transgenic farm animals that secrete medically useful products in their urine, which has commercial advantages over secretion in milk.

Moffat, Anne Simon. "Exploring Transgenic Plants as a New Vaccine Source." *Science*, vol. 268, May 5, 1995, pp. 658-659. Transgenic plants may allow scientists to produce vaccines that can be eaten, which would be cheaper, safer, and easier to store and administer than conventional ones and thus might be especially useful in developing countries.

———. "High-Tech Plants Promise a Bumper Crop of New Products." *Science*, vol. 256, May 8, 1992, pp. 770–771. In the early 1990s, researchers became able to genetically engineer common crop plants to manufacture a wide variety of materials, including human proteins useful in medicine, a bacterial enzyme used in the food processing industry, and natural polymers such as a type of polyester.

Morell, Virginia. "A Clone of One's Own." *Discover*, vol. 19, May 1998, pp. 83–89. Focuses on the cloning of monkeys from embryo cells in an Oregon laboratory. Also describes other experimental cloning of mammals and considers ethical implications of cloning humans.

Nash, J. Madeleine. "The Age of Cloning." *Time*, vol. 149, March 10, 1997, pp. 62–66. The cloning of Dolly the sheep from a body cell of an adult ewe, described here, raises interesting scientific questions about the function and maturation of cells and ethical issues as well.

Nestle, Marion. "Allergies to Transgenic Foods—Questions of Policy." *New England Journal of Medicine*, vol. 334, March 14, 1996, pp. 726ff. Genetic engineering of food plants may carry allergenic properties of the donor into the recipient plant, as happened with a Brazil nut gene inserted recently into soy plants. Author maintains that present FDA regulations are inadequate to protect consumers against the resulting allergy risk from transgenic foods.

Office of Science and Technology Policy. "Coordinated Framework for Regulation of Biotechnology." *Federal Register*, vol. 51, 1986, pp. 23302–23393. Allocates safety regulation of biotechnology among five different government agencies and spells out the responsibilities of each.

———. "Exercise of Federal Oversight Within the Scope of Statutory Authority: Planned Introductions of Biotechnology Products into the Environment." *Federal Register*, vol. 57, 1992, pp. 6753–6762. Describes scope of regulations concerning genetically modified organisms and states that these should be made on the basis of the characteristics of the organisms rather than on the fact that they were genetically engineered.

Paoletti, Maurizio G., and David Pimentel. "Genetic Engineering in Agriculture and the Environment: Assessing Risks and Benefits." *BioScience*, vol. 46, October 1996, pp. 665ff. Genetic engineering of farm plants and animals offers the possibility of increasing productivity and resistance to disease, but the introduction of genetically engineered living things into the environment must be carefully regulated.

Pellegrino, Charles. "Resurrecting Dinosaurs." *Omni*, vol. 17, Fall 1995, pp. 68ff. Some scientists think that the technology needed to clone dinosaurs from fragments of DNA such as those found in amber will be available in about 20 years.

Pennisi, Elizabeth. "After Dolly, a Pharming Frenzy." *Science*, vol. 279, January 30, 1998, pp. 646ff. The cloning techniques that created Dolly the sheep and two calf clones are being geared up to produce herds of transgenic animals that could make substances useful in medicine or research, but the technology is still very inefficient and needs much improvement.

———. "Cloned Mice Provide Company for Dolly." *Science*, vol. 281, July 24, 1998, pp. 495–496. Advances in animal cloning in the wake of Dolly include production of more than 50 mice cloned from adult cells at the University of Hawaii; confirmation that Dolly and the ewe she was cloned from are genetically identical; and creation of two similarly identical calves in Japan.

Pennisi, Elizabeth, and Nigel Williams. "Will Dolly Send in the Clones?" *Science*, vol. 275, March 7, 1997, pp. 1415–1416. Points out the scientific questions yet to be answered about cloning in the wake of Ian Wilmut's cloning of a sheep from an adult sheep's body cell; sidebar considers possible legal ramifications of cloning.

Annotated Bibliography

Perkins, Sid. "Transgenic Plants Provoke Petition." *Science News*, vol. 152, September 27, 1997, p.199. More than 20 groups, including environmentalists and organic farmers, filed a petition asking the Environmental Protection Agency to rescind approvals of a group of plants genetically engineered to produce a pesticide because they feared that steady production of the substance would encourage insects to develop resistance to it.

Powledge, Tabitha M. "Goodbye, Dolly." *Technology Review*, vol. 100, May–June 1997, p. 5. Opposes a potential ban on cloning research because cloning could be used to save endangered species, increase the productivity of livestock, and improve drug research.

Praded, Joni. "Cloning: The Missing Debate." *Animals*, vol. 130, May–June 1997, pp. 21ff. Although the ethics of possible human cloning has been much discussed, the ethics of animal cloning has not; author wants a focus on cultural values in developing policies on application of cloning and other genetic engineering techniques to animals.

Prakash, C. S. "Monsanto Will Wait for Studies of Disputed New Gene Technology." *St. Louis Post-Dispatch*, April 23, 1999. Bowing to widespread controversy, Monsanto Corporation has decided to hold off marketing crops that contain inserted genes that keep them from reproducing (called "Terminator" genes by opponents) until studies examining the environmental, economic, and social effects of such crops can be completed.

Raloff, Janet. "Ferreting Out Cancer Risk with Novel Mice." *Science News*, vol. 148, October 21, 1995, p. 263. Two types of genetically altered mice seem to be able to screen chemicals for carcinogenicity more cheaply and effectively than current tests can; one carries only a single copy of a cancer-suppressing gene, while the other has a cancer-causing gene that is activated by carcinogens.

"Report of the Joint EPA FIFRA Scientific Advisory Panel and Biotechnology Science Advisory Committee Subpanel on Plant Pesticides." *Federal Register*, vol. 59, February 10, 1994. Concludes that plants to which genes for toxin from bacterium *Bacillus thuringensis* (Bt) have been added are pesticides and therefore are subject to EPA review under the Federal Insecticide, Fungicide, and Rodenticide Act (FIFRA).

Rifkin, Jeremy. "Dolly's Legacy: The Implications of Animal Cloning." *The Animals' Agenda*, vol. 17, May–June 1997, pp. 34–35. Rifkin, a frequent critic of biotechnology, describes the questions about animal rights raised by the cloning of a sheep; he calls animal cloning "the most fundamental violation of animal rights in history."

———. "Future Pharming." *Animals*, vol. 131, May–June 1998, pp. 24ff. Creation of transgenic animals may advance medical and agricultural technologies, but great potential for abusing the animals also exists;

humans should reexamine their relationship with their fellow species before proceeding with the biotechnology revolution.

Rippel, Barbara. "Feeding the Future: Genetically Modified Food." *Consumers' Research Magazine*, vol. 81, September 1998, pp. 34–35. Americans appear more willing to accept genetically engineered foods than Europeans. Differences in policy about such foods could affect trade relations between the European Union and the United States.

Rosenfeld, Albert. "New Breeds Down on the Pharm." *Smithsonian*, vol. 29, July 1998, pp. 23–30. Describes transgenic farm animals that contain human genes and are thus potential factories for rare hormones or other medically useful substances, including possibly tissues or organs for transplanting.

Rotman, David. "The Next Biotech Harvest." *Technology Review*, vol. 101, September–October 1998, pp. 34ff. Chemical and agricultural companies are merging with biotechnology firms to take advantage of a predicted boom in biotech that is expected to revolutionize the American agricultural industry.

Sardar, Ziauddin. "Loss of Innocence." *New Statesman*, vol. 129, February 26, 1999, pp. 47–48. Describes alleged suppression of a report by a British scientist, Arpad Pusztai, who found liver damage and other health problems in animals fed genetically engineered potatoes. Author says Pusztai's case shows that scientists can no longer be assumed to be objective and free from ties to or coercions from industry.

Seiya, Takahashi. "Until the Cows Come Cloned." *Look Japan*, January 1999, pp. 28–29. In July 1998, Tsunoda Yukio of Kinki University in Japan produced the world's first calves cloned from a body cell of a mature cow. The article describes this and other Japanese advances in animal cloning technology.

Shapiro, Robert B. "How Genetic Engineering Will Save Our Planet." *The Futurist*, vol. 33, April 1999, pp. 28–29. Author, the CEO of Monsanto Corporation, says that agricultural methods in the 20th century have not been sustainable, but biotechnology and information technology could allow sustainability to be achieved in future.

Shonsey, Edward T. "Biotechnology and Transgenic Products: Evolution of Global Markets." *Vital Speeches*, vol. 65, March 15, 1999, pp. 342ff. Shonsey, president and CEO of Novartis Seeds, Inc., says that *evolution* is a better term than *revolution* to explain biotechnology's past, present, and future progress.

Shuldiner, Alan R. "Transgenic Animals." *New England Journal of Medicine*, vol. 334, March 7, 1996, pp. 653–654. Describes use of transgenic animals, especially mice, as experimental models of human disease. Explains the techniques used to produce the animals.

Annotated Bibliography

Snow, Allison A., and Pedro Moran Palma. "Commercialization of Transgenic Plants." *Bioscience*, vol. 47, February 1997, pp. 86ff. Warns that large-scale use of transgenic plants may pose environmental problems such as increases in pest resistance to insecticides and changes in soil fertility.

Specter, Michael, with Gina Kolata. "After Decades and Many Missteps, Cloning Success." *New York Times*, March 3, pp. 1997, A-1, A-8–A-10. Newspaper story describes the production of a lamb (Dolly) from a cloned udder cell of an ewe by Ian Wilmut and others at the Roslin Institute in Scotland, the first cloning of a fully developed mammal from a mature body cell of an adult animal.

Spencer, Peter. "No Pipe Dream." *Consumer's Research Magazine*, vol. 82, June 1999, p. 43. Describes current and coming generations of bioengineered crops and their benefits to farmers and consumers.

Stone, Richard. "Cloning the Woolly Mammoth." *Discover*, vol. 20, April 1999, pp. 56–63. Japanese scientists hope to clone a woolly mammoth from recovered bits of the extinct animal's DNA.

Straughan, R. "Ethics, Morality, and Crop Biotechnology: Extrinsic Concerns About Consequences." *Outlook on Agriculture*, vol. 24, December 1995, pp. 233–240. Considers issues in the ethics of agricultural biotechnology, including risks of genetically engineered crops to the environment and human health and the economic and social impact of biotechnology patents.

Tanksley, Steven D., and Susan R. McCouch. "Seed Banking and Molecular Maps: Unlocking Genetic Potential from the Wild," *Science*, vol. 277, August 22, 1997, pp. 1063ff. The usefulness of existing plant germplasm (seed) banks can be greatly increased by the new technique of genetic linkage mapping, which allows science to pinpoint the chromosomal location of genes that govern traits valuable in agriculture.

Travis, John. "A Fantastical Experiment: The Science Behind the Controversial Cloning of Dolly." *Science News*, vol. 151, April 5, 1997, pp. 214–215. Describes the scientific background and importance of Ian Wilmut's cloning of a lamb from a mature udder cell of an adult ewe.

United States Department of Agriculture. "Plant Pests: Introduction of Genetically Engineered Organisms or Products." *Federal Register*, vol. 52, 1987, pp. 22892–22915. Describes new regulations covering transgenic plants.

Velander, William H., Henryk Lubon, and William N. Drohan. "Transgenic Livestock as Drug Factories." *Scientific American*, vol. 276, January 1997, pp. 70–74. Biologists are introducing human genes into cattle and other large mammals to make them produce medically useful proteins in their milk; this form of production is both cheaper and less open to contamination than conventional methods.

Vrijenhoek, Robert C. "Animal Clones and Diversity." *BioScience*, vol. 48, August 1998, p. 617. Clones are sometimes produced in nature through asexual reproduction, but such clones are more likely to be the products of mutation and usually have less adaptive potential than their normal counterparts.

Williams, Nigel. "Agricultural Biotech Faces Backlash in Europe." *Science*, vol. 281, August 7, 1998, pp. 768ff. Genetically modified foods have faced relatively little consumer resistance in the United States, but Europeans are expressing fears about possible dangers of such foods to both humans and the environment.

Wills, Christopher. "A Sheep in Sheep's Clothing?" *Discover*, vol. 18, January 1998, pp. 22–23. Using Dolly, the famous cloned sheep, as an example, author illustrates the usefulness and ethical viability of cloning in animals and explains why human cloning might also be ethical under some circumstances.

Wilmut, I., et al. "Viable Offspring Derived from Fetal and Adult Mammalian Cells." *Nature*, vol. 385, February 27, 1997, pp. 810–813. This scientific paper describes the creation of Dolly the sheep, the first fully developed mammal created by cloning a mature body cell from an adult animal.

Yap, Wendy, and David Rejeski. "Environmental Policy in the Age of Genetics." *Issues in Science and Technology*, vol. 15, Fall 1998, pp. 33ff. Although the new technology of gene chips offers the potential for preventing or minimizing human exposure to environmental threats at a molecular level, it can be misused in ways that produce institutional and moral disaster.

INTERNET/WEB DOCUMENTS

Kayotic Development. "The Three Ways to Clone Mammals." (1998) Human Cloning Foundation website. Available online. URL: http://www.humancloning.org/threeways.htm. Posted March 3, 1999.

National Institutes of Health. *Guidelines for Research on Recombinant DNA Molecules*. Available online. URL: http://www.nih.gov/od/oba/guidelines.htm. Updated May 1999. First promulgated in 1976, these guidelines have been updated to keep in step with present DNA research. They classify experiments into different risk groups and list safety and other regulations applicable to each.

Nuffield Council on Bioethics. *Genetically Modified Crops: The Ethical and Social Issues*. NCBE/University of Reading web site. Available online. URL: http://134.225.167.114/NCBE/GMFOOD/menu.html. Posted May 27, 1999. Reviews genetic modification of crops and its ethical and social implications, including issues of food safety, public health, environmental protection, intellectual property rights, and government policy and regulation. Makes recommendations.

Trebach, Susan, et al. "Cloning: Hello Dolly!" Available online. URL: http://whyfiles.news.wisc.edu/034clone/index.html. Posted March 6, 1997. Part of the Why Files, a project funded by the National Science Foundation, this online article provides a lively introduction to the science and ethics of cloning, both animal and human.

Trebach, Susan, et al. "Field of Genes: Agricultural Genetic Engineering." Available online. URL: http://whyfiles.news.wisc.edu/062ag_gene_eng/index.html. Posted April 23, 1998. Part of the Why Files, a project funded by the National Science Foundation, this online article provides a lively introduction to the science and ethics of genetically engineered crops.

PATENTING LIFE

BOOKS

Burchfield, K. J. *Biotechnology and the Federal Circuit.* Washington, D.C.: BNA Books, 1995. Uses cases reviewed by the United States Federal Circuit Courts of Appeals to describe recent changes in United States patent laws affecting biotechnology.

Cole-Turner, R. *The New Genesis: Theology and the Genetic Revolution.* Louisville, Ky.: Westminster John Knox, 1993. Maintains that there is no distinctly religious ground for opposing the patenting of DNA sequences.

Office of Technology Assessment. *New Developments in Biotechnology: Ownership of Human Tissues and Cells.* Washington, D.C.: Office of Technology Assessment, 1987. Considers legal status of patents on tissues and cells and the question of whether the originators of those tissues and cells have any ownership rights to them.

Sibley, K. D. *The Law and Strategy of Biotechnology Patents.* Boston, Mass.: Butterworth-Heinemann, 1994. Describes current patent law in the United States as it applies to biotechnology and considers how the industry can best use patents.

Wijk, J. V., J. I. Cohen, and J. Komen. *Intellectual Property Rights for Agricultural Biotechnology: Options and Implications for Developing Countries.* The Hague, the Netherlands: International Service for National Agricultural Research, 1993. Describes laws and legislation related to agricultural biotechnology in developing countries, including protection of intellectual property rights.

Biotechnology and Genetic Engineering

World Resources Institute et al. *Global Biodiversity Strategy.* Washington, D.C.: World Resources Institute, International Union for the Conservation of Nature, United Nations Environment Program, 1992. Presents a strategy for gathering useful genetic information from the planet's biodiversity while at the same time preventing abuse and encouraging rural development.

ARTICLES

Adler, Reid G. "Biotechnology as an Intellectual Property." *Science*, vol. 224, April 27, 1984, pp. 357ff. Provides a detailed history of patenting and other ways of protecting intellectual property as they apply to biotechnology, including comparison of various statutes.

Armstrong, Jeannette. "Global Trade Targets Indigenous Gene Lines." *National Catholic Reporter*, vol. 32, January 27, 1995, pp. 11–12. Maintains that trade treaties such as NAFTA and GATT erode the protection of indigenous peoples and their lands, leaving them vulnerable to bioprospectors who seek genetic information for large biotechnology companies to exploit.

Barton, John H. "Adapting the Intellectual Property System to New Technologies." *International Journal of Technology Management*, vol. 10, February–March 1995, pp. 151–72. Considers how to adapt the patent system and other means of protecting intellectual property to new industries, including biotechnology.

———. "Patenting Life." *Scientific American*, vol. 264, March 1991, pp. 40–46. Describes some of the chief legal conflicts related to the biotechnology industry's requests for patents on living things.

Brownlee, Shannon. "Staking Claims on the Human Body." *U.S. News & World Report*, vol. 111, November 18, 1991, p. 89. The biotechnology company SyStemix was granted a patent on use of stem cells to treat disease. This is the first patent issued for a part of the human body, and it raises controversial legal and ethical issues.

Curran, William J. "Scientific and Commercial Development of Human Cell Lines: Issues of Property, Ethics, and Conflict of Interest." *New England Journal of Medicine*, vol. 324, April 4, 1991, pp. 998ff. Discusses the implications of the California Supreme Court ruling in *Moore v. Regents of the University of California*.

Dolnick, Edward. "Spare Parts: Whose Body Is It, Anyway?" *New Republic*, vol. 195, September 15, 1986, pp. 16–17. Describes the court case involving businessman John Moore, who sued his doctor and others who had patented a cell line derived from his surgically removed spleen.

170

Eisenberg, Rebecca. "Genes, Patents, and Product Development." *Science*, vol. 257, August 14, 1992, pp. 903–908. Discusses the negative reaction of industry trade groups to NIH's attempt to patent thousands of DNA sequences of unknown function.

———. "A Technological Policy Perspective on the NIH Gene Patent Controversy." *University of Pittsburgh Law Review*, vol. 55, 1994, pp. 633–652. Uses this issue as a focus for evaluating the role of patents in publicly funded biotechnology research, including the Human Genome Project.

Feldbaum, Carl B. "Should the Government Prevent the Patenting of Genetically Engineered Animal and Human Genes, Cell Lines, Tissues, Organs and Embryos?" *CQ Researcher*, vol. 5, December 8, 1995, p. 1105. Presents two opposing viewpoints on the question.

Gillis, Anna Maria. "The Patent Question of the Year." *BioScience*, vol. 42, May 1992, pp. 336ff. Describes the debate over attempts by the National Institutes of Health to patent fragments of human brain DNA and the implications of this issue for the biotechnology industry.

Gross, Neil, and John Carey. "Who Owns the Tree of Life?" *Business Week*, no. 3500, November 4, 1996, pp. 194ff. There is considerable legal and ethical controversy over the patenting of genes, which the biotechnology industry says is necessary to protect various types of research, including cancer research. Opponents, however, claim that people cannot or should not hold a patent on life.

Harness, Charles L. "The Bug, the Mouse, and Chapter 24." *Analog Science Fiction and Fact*, vol. 114, December 15, 1994, pp. 70ff. Describes a 1993 addition to the U.S. Patent and Trademark Office's Manual of Patent Examining Procedure concerned with genetic engineering and biotechnology (Chapter 24), including the reasons and court cases behind the move.

Horvitz, Leslie Alan. " 'Vampire Project' Raises Issue of Patents for Human Genes." *Insight on the News*, vol. 12, July 22, 1996, pp. 34–35. Tribal rights groups fear that the Human Genome Diversity Project, which intends to gather genetic information about indigenous peoples through samples of hair and blood, may lead to the "stealing" of this information for the benefit of large drug or biotechnology companies.

Kaiser, Jocelyn. "New Biotech Law Shores up U.S. Firms." *Science*, vol. 270, November 3, 1995, p. 728. A new law makes familiar genetic engineering processes patentable if they use or produce a novel product.

Kiley, Thomas D. "Patents on Random Complementary DNA Fragments?" *Science*, vol. 257, August 14, 1992, pp. 915ff. The proposal by the National Institutes of Health to patent sequences from the human genome is wrong in patent law or, at best, relies on deficiencies in the law that need to be corrected so that the raw material of scientific experimenting (as opposed to "useful" inventions made from it) cannot be owned.

Kimbrell, Andrew. "High-tech Piracy." *Utne Reader*, no. 74, March–April 1996, pp. 84ff. Because life forms and genetic sequences can be patented, private companies may be able to exploit genetic information taken from the environment or even the bodies of indigenous peoples unless international structures are put in place to protect the genetic integrity of life.

Marsa, Linda. "Whose Ideas Are They, Anyway? Intellectual Property in the Information Age." *Omni*, vol. 17, Winter 1995, pp. 36ff. Includes a discussion of the patenting of genetic discoveries.

Marshall, Eliot. "The Company that Genome Researchers Love to Hate." *Science*, vol. 266, December 16, 1994, pp. 1800ff. Human Genome Sciences, Inc.(HGS) and its nonprofit partner, The Institute for Genomic Research (TIGR), are doing some of the same type of research as the government-funded Human Genome Project, but they are patenting the results, which has led to considerable controversy.

———. "A Showdown over Gene Fragments." *Science*, vol. 222, October 14, 1994, pp. 208-210. Genome researchers are upset because private companies' patenting of DNA sequences, such as the recently discovered BRCA1 gene, threaten to create barriers to the free exchange of information. They say such data should be in a public data bank.

———. "Whose DNA Is It, Anyway?" *Science*, vol. 278, October 24, 1997, pp. 564ff. Describes conflict between the National Institutes of Health, which is trying to open up access to DNA data from projects it has funded, and researchers and biotechnology companies, who want to keep control of the information they discovered.

Matthews, K. I. "Recent Developments in Animal Patenting." *Genewatch*, vol. 8, March 1993, p. 5. Describes developments in the patenting of genetically engineered livestock and laboratory animals (mice).

Menon, U. "Access to and Transfer of Genetic Resources." *International Journal of Technology Management*, vol. 10, February–March 1995, pp. 311–324. Considers methods of technology transfer, regulations protecting intellectual property, and agreements related to outside development of agricultural and other genetic resources in such countries as India and Costa Rica.

Newman, S., and N. Wilker. "The CRG Says No to Patenting Life Forms." *Genewatch*, vol. 8, July 1992, pp. 8–9. The Council for Responsible Genetics opposes the patenting of transgenic plants and animals.

O'Brien, Claire. "European Parliament Axes Patent Policy." *Science*, vol. 267, March 10, 1995, pp. 1417–1418. The European Parliament voted down proposed standards for biotechnology patents that were supported by genetic researchers but opposed by environmentalists and animal rights groups, who do not wish to see living things patented.

"Patents for Seeds and Plants." *Science News*, vol. 128, October 26, 1985, p. 267. Plants, seeds, and plant tissue cultures can now be patented, the U.S. Board of Patent Appeals and Interferences has ruled; formerly, only plant varieties reproducing asexually and single, novel varieties of sexually reproducing plants could be patented.

Powledge, Fred. "Who Owns Rice and Beans?" *Bioscience*, vol. 45, July–August 1995, pp. 440ff. The U.S. Patent and Trademark Office has had confusing and inconsistent policies about patenting plant germplasm, for instance granting broad patents to Agracetus in 1991–1992 and then nullifying them after the biotechnology community protested.

Raines, L. J. "Protecting Biotechnology's Pioneers." *Issues in Science and Technology*, vol. 8, Winter 1991–1992, pp. 33–39. Describes some patents held or applied for by the scientists who founded genetic engineering.

Raustiala, K., and D. G. Victor. "Biodiversity Since Rio: The Future of the Convention on Biological Diversity." *Environment*, vol. 38, May 1996, pp. 17–20, 37–45. Considers how the agreement made at the Rio Earth Summit will be affected by the conflict between the attempt to preserve biodiversity and the biotechnology industry's drive to protect its intellectual property rights.

Rood, M. "Animal Welfare Groups Battle Harvard's Onco-mouse Patent." *Biotech Daily*, vol. 2, January 1993, pp. 1, 3. Animal welfare groups oppose the granting of the first patent on a higher animal.

Rudolph, J. R. "Patentable Invention in Biotechnology." *Biotechnology Advances*, vol. 14, 1996, pp. 17–34. Considers what makes a biotechnology invention patentable and what rights a biotechnology patent confers.

Sagoff, Mark. "Patented Genes: An Ethical Appraisal." *Issues in Science and Technology*, vol. 14, Spring 1998, pp. 37ff. Religious leaders from 80 faiths have presented a "Joint Appeal Against Human and Animal Patenting" in the hope of reversing the decision of the U.S. Patent Office to permit patenting of body parts and genetically engineered animals. Their efforts worry the biotechnology industry, but reconciliation of the two groups' positions may be possible.

Sedjo, Roger A. "Property Rights, Genetic Resources, and Biotechnological Change." *Journal of Law and Economics*, vol. 35, April 1992, pp. 199–213. Discusses the potential conflicts between the intellectual property rights of biotechnology companies and indigenous people's rights to their land's genetic resources that are created by bioprospecting.

Shreeve, James. "The Code Breaker." *Discover*, vol. 19, May 1998, pp. 44–51. Craig Venter, who heads the Institute for Genomic Research (TIGR) in Rockville, Maryland, worked out a process for sequencing

DNA very rapidly and thereby decoding the genomes of a variety of living things, including humans. He soon found himself embroiled in a controversy about whether this information could be patented.

Stone, Richard. "Sweeping Patents Put Biotech Companies on Warpath." *Science*, vol. 268, May 5, 1995, pp. 656–658. Biotechnology firms are challenging four genetic engineering patents issued since 1992, including two given to Agracetus for all genetically altered cotton and soybeans, because they claim the patents are too broad in scope.

Strauss, Evelyn. "The Tissue Issue." *Science News*, vol. 152, September 20, 1997, pp. 190–191. Describes the debate over who owns cells or tissues once they have been removed from a person's body and who controls their use in research and potential profit-making ventures.

Tangley, Laura. "Who Owns Human Tissues and Cells?" *BioScience*, vol. 37, June 1987, pp. 376ff. Summarizes an Office of Technology Assessment report on questions concerning the patenting of human tissues and cells used in biomedical research; considers possible legislation on this subject.

Taubes, Gary. "Scientists Attacked for 'Patenting' Pacific Tribe." *Science*, vol. 270, November 17, 1995, p. 1112. A patent on a cell line made from the blood of a member of the Hagahai people of Papua New Guinea has attracted media criticism, partly because it has been misrepresented as being a patent on a human being.

Thompson, Paul B. "Conceptions of Property and the Biotechnology Debate." *BioScience*, vol. 45, April 1995, pp. 275ff. Presents a detailed analysis of the law and philosophy of property as it relates to the idea of patenting plants, animals, genes, and processes in living bodies.

———. "Designing Animals: Ethical Issues for Genetic Engineers." *Journal of Dairy Science*, vol. 75, August 1992, pp. 2294–2303. Considers two philosophical approaches to intellectual property rights, the instrumental approach and the labor approach, and describes the application of these approaches to the ethics of animal biotechnology. Includes discussion of ethical issues raised by unwanted consequences of biotechnology and by religious objections to genetic alteration.

Tsevdos, Estelle, Robin A. Chadwick, and Gail Matthews. "Law and Nature Collide." *National Law Journal*, June 16, 1997, pp. C1, C27, C29–C30. Discusses whether cloned animals such as the famous sheep, Dolly, are patentable and concludes that the possibility is "iffy."

Vogel, Shawna. "Patented Animals." *Technology Review*, vol. 91, October 1988, pp. 15–16. In 1988, after a mouse genetically engineered for use in tests of potential carcinogens became the first animal to be patented, some legislators called for a moratorium on the patenting of animals.

Williams, Nigel. "European Parliament Backs New Biopatent Guidelines." *Science*, vol. 275, July 25, 1997, p. 472. After an intense lobbying campaign, members of the European Parliament approved an outline of legislation that will determine which biotechnology inventions can be patented in the European Union.

Wright, Karen. "The Body Bazaar." *Discover*, vol. 19, October 1998, pp. 115–120. Joint ventures between biotechnology companies and hospitals with tissue banks bring up questions about who owns cells and tissues removed from people's bodies.

INTERNET/WEB DOCUMENTS

Dobert, Raymond. *Biotechnology: Patenting Issues: January 1990–July 1996.* Quick Bibliography Series 96–09. Biotechnology Information Center, National Agricultural Library, Agricultural Research Service, U.S. Department of Agriculture. Available online. URL: http://www.nal.usda. gov/bic/Biblios/patentag.htm. Posted September 1996.

Elman, Gerry J. *Mammalian Cloning: Crosscurrents in the Patent System.* Available online. URL: http://www.canniff.net/law/talks/elman/19970627/. Posted July 3, 1997. Provides history of laws and biotechnology developments that affect patenting of genetically altered animals and cloned animals, considers reasons for opposing the patenting of animals, and offers examples of patents pertaining to genetically altered or cloned animals.

OTHER MEDIA

Biotechnology Patents: Documentation to Support Patents. Two videocassettes, 171 minutes. Potomac, Md.: BioConferences International, 1991. Videorecording of a session in BioEast '91 conference (held in Washington, D.C., January 6–9, 1991) in which a panel discussed national and international requirements for biotechnology patents, uses of patents to protect market areas, and the judicial climate regarding biotechnology and patents.

HUMAN GENETICS

BOOKS

Annas, George J., and Sherman Elias, eds. *Gene Mapping: Using Law and Ethics as Guides.* New York: Oxford University Press, 1992. Essay collection discusses the Human Genome Project and the social policy issues it raises, including those of eugenics, genetic discrimination, medical privacy, gene patenting, genetic determinism, and gene therapy.

Appleyard, Bryan. *Brave New Worlds: Staying Human in the Genetic Future.* New York: Viking, 1998. Author warns that the science of human genetics is advancing quickly and that people must not expect scientists to solve the ethical dilemmas it presents.

Cohen, David B. *Strangers in the Nest: Do Parents Really Shape Their Child's Personality, Intelligence, or Character?* New York: Wiley, 1999. Maintains that genes have more influence than parental environment on children's development.

Cook-Deegan, R. *The Gene Wars.* New York: Norton, 1994. Describes early stages and conflicts in the Human Genome Project; includes discussion of the project's ethical, legal, and social implications.

Drlica, Karl A. *Double-Edged Sword: The Promises and Risks of the Genetic Revolution.* Reading, Mass.: Addison-Wesley, 1994. Focuses primarily on possible uses and misuses of human genetic information in medicine, including ethical issues; has chapters on DNA fingerprinting and genetic testing.

Hamer, Dean, and Peter Copeland. *Living with Our Genes: Why They Matter More than You Think.* Garden City, N.Y.: Doubleday, 1998. Describes what is currently known about the role of genetics in shaping human behavior. Claims that genes have a large but by no means exclusive role in producing such behaviors as thrill seeking, drug abuse, and violence.

Harris, John. *Clones, Genes, and Immortality: Ethics and the Genetic Revolution.* New York: Oxford University Press, 1998. Discusses ethical dilemmas involved in research on human embryos, commercialization of human genetic information, alteration of human genes, and possible discrimination resulting from genetic testing.

Heller, Jan Christian. *Human Genome Research and the Challenge of Contingent Future Persons: Toward an Impersonal Theocentric Approach to Value.* Omaha, Neb.: Creighton University Press, 1996. Discusses the implications of the Human Genome Project and related research for future generations from a theological perspective.

House of Commons Science and Technology Committee. *Human Genetics: The Science and Its Consequences.* London: HM Stationery Office, 1995. British government committee report considers positive and negative consequences of the increased knowledge gained through the Human Genome Project and related research, including the possibility of discrimination based on genetic profiles.

Hubbard, Ruth, and Elijah Wald. *Exploding the Gene Myth: How Genetic Information Is Produced and Manipulated by Scientists, Physicians, Employers, Insurance Companies, Educators, and Law Enforcers.* Boston: Beacon Press, 1997. Thoughtful book on the downside of production and use of information from the human genome focuses on genetic testing and possible misuses of genetic information, especially in a medical context. It also has chapters on patents/conflict of interest and DNA identification/testing.

Jones, Steve. *The Language of Genes: Solving the Mysteries of Our Genetic Past, Present, and Future.* Garden City, N.Y.: Doubleday, 1995. Good introduction to the Human Genome Project and other recent research on human genetics and its social and ethical implications, such as the possibility of a return to eugenics; corrects common misconceptions spread by the media.

Kevles, Daniel J., and Leroy Hood, eds. *The Code of Codes: Scientific and Social Issues in the Human Genome Project.* Cambridge, Mass.: Harvard University Press, 1992. Still an excellent source in spite of its early publication date, this anthology includes articles on such subjects as the "historical politics of the human genome," identification through DNA, genetic screening, and gene therapy.

Kidd, J. S., and Renee A. Kidd. *Life Lines: The Story of the New Genetics.* New York: Facts On File, 1999. Book for young adults in the Science and Society series provides background in both classical genetics and recent discoveries, especially in human genetics, and goes on to discuss such subjects as genetic testing for disease, gene therapy, the use of genetics in crime detection, and cloning.

Kilner, John F., Rebecca D. Pentz, and Frank E. Young, eds. *Genetic Ethics: Do the Ends Justify the Genes?* Grand Rapids, Mich.: William B. Eerdmans, 1997. Collection of essays by scholars and scientists in the forefront of genetic research considers from a Christian perspective the ethical dilemmas being raised by advances in human genetics, including use of genetic information and possible changing of human genes.

Kitcher, Philip. *The Lives to Come: The Genetic Revolution and Human Possibilities.* New York: Simon and Schuster, 1996. Describes human genome research, its likely near-term contributions to medicine, and the moral and social issues it raises. The author, an advocate of "utopian genetics," supports abortion of fetuses with serious genetic abnormalities but not alteration to produce "ideal" offspring.

Lewontin, Richard. *Human Diversity.* New York: W. H. Freeman, 1995. Famed Harvard University population biologist explains the genetic basis of human diversity and warns against assuming that genes are the chief determiners of human personality and behavior or of differences between races.

Marshall, Elizabeth L. *The Human Genome Project: Cracking the Code Within Us.* Danbury, Conn.: Franklin Watts, 1997. Book for young adults describes the Human Genome Project, using the research of individual scientists as examples to clarify the project's basic nature. In discussing ethical issues, the author stresses the project's medical benefits.

Marteau, Theresa, and Martin Richards, eds. *The Troubled Helix: Social and*

Psychological Implications of the New Genetics. New York: Cambridge University Press, 1996. Surveys opportunities and problems stemming from new discoveries about human genetics, such as the fact that genetic tests often can detect diseases for which no treatment or cure is available.

Mehlman, Maxwell A., and Jeffrey R. Botkin. *Access to the Genome: The Challenge to Equality*. Washington, D.C.: Georgetown University Press, 1998. Describes issues raised by the Human Genome Project and related research, including genetic testing and screening, gene therapy, and genetic enhancement; considers problems with unequal access to genetic technologies and possible ways of making access more fair.

MITRE Corporation. *JASON Report JSR-97–315*. Maclean, Va.: MITRE Corporation, 1997. Major federal evaluation of the Human Genome Project suggests that the project has serious problems, including slow development of DNA sequencing technology and lack of standardization in data reporting, software tools, and terminology.

National Research Council. *Evaluating Human Genetic Diversity*. Washington, D.C.: National Academy Press, 1998. Discusses the scientific and medical value of research on human genetic variation as well as issues that genetic diversity projects raise about sampling and human rights.

Nelkin, Dorothy, and Susan M. Lindee. *The DNA Mystique: The Gene as a Cultural Icon*. New York: W. H. Freeman, 1996. Describes popular culture's fascination with genes as a way to explain human behavior and considers the effects of this genetic determinism on social policy, individual expectations, and reproductive practices.

Neubauer, Peter B., and Alexander Neubauer. *Nature's Thumbprint: The New Genetics of Personality*. New York: Columbia University Press, 1996. Insists that nature and nurture (genetics and environment) are equally important in shaping human personality and complex behaviors.

O'Neil, Terry. *Biomedical Ethics: Opposing Viewpoints Digest*. San Diego: Greenhaven Press, 1999. Summary of opposing viewpoints on several subjects in biomedical ethics includes chapters on genetic testing and therapy and on human cloning.

Paul, Diane B. *The Politics of Heredity: Essays on Eugenics, Biomedicine, and the Nature-Nurture Debate*. Albany, N.Y.: State University of New York Press, 1998. Essays examine the political factors underlying changes in beliefs about the relative importance of nature (genes) and nurture (environment) in shaping human behavior and personality.

Peters, Ted, ed. *Genetics: Issues of Social Justice*. Cleveland, Ohio: Pilgrim Press, 1998. Collection of essays by leading ethicists, theologians, scientists, and legal experts discuss the social and religious implications of recent advances in research on human genetics.

————. *Playing God? Genetic Determinism and Human Freedom*. New York: Routledge, 1997. Describes the growing acceptance of genetic determinism, the belief that genes control such complex behaviors as homosexuality and violence. Warns that this belief can lead to discrimination and undesirable alteration of the human germ line, limiting the freedom not only of individuals but of all their descendants.

Pierce, Benjamin A. *The Family Genetic Sourcebook*. New York: Wiley, 1990. Intended for the layman, this book provides a catalog of inherited traits, including those related to various illnesses, as well as a concise description of heredity and reproduction.

Roleff, Tamara L., ed. *Genetics and Intelligence*. San Diego: Greenhaven Press, 1996. Anthology of articles includes material on whether genetics influences intelligence and whether genetics is responsible for differences between races in IQ scores.

Rothman, Barbara Katz. *Genetic Maps and Human Imaginations: The Limits of Science in Understanding Who We Are*. New York: W. W. Norton, 1998. Author, a sociologist, uses personal experience as a springboard for exploring the social implications of recent advances in human genetics. She attacks the Human Genome Project and belief in genetic determinism and expresses fear that genetic advances will lead to a new racism, turn beauty and intelligence into products for sale, or limit toleration of difference and individuality.

Steen, R. Grant. *DNA and Destiny: Nature and Nurture in Human Behavior*. New York: Plenum Press, 1996. Author takes on the question of whether nature or nurture—genes or environment—determine such complex characteristics as intelligence or a tendency to violence and presents evidence demonstrating subtle interactions between the two.

Thomasma, David C., and Thomasine Kushner, eds. *Birth to Death: Science and Bioethics*. New York: Cambridge University Press, 1996. Reviews major recent advances in biology, especially genetics, and considers their ethical implications.

Weir, Robert F. *Genes and Human Self-Knowledge: Historical and Philosophical Reflections on Modern Genetics*. Iowa City: University of Iowa Press, 1994. Collection of essays considers such topics as the effect of new knowledge from the Human Genome Project on people's feeling of identity and self-knowledge, whether advances in human genetics will lead to discrimination or a new eugenics, how the media's portrayal of genetics affects public views on the subject, and the ethics of changing human germ-line genes.

Wills, Christopher. *Children of Prometheus: The Accelerating Pace of Human Evolution*. Reading, Mass.: Perseus, 1998. Author, a British evolutionary biologist, explains his belief that the evolution of humans is speeding up

because of changes that people are making in their environment. He predicts an optimistic future for the species.

Wingerson, Lois. *Unnatural Selection: The Promise and the Power of Human
Gene Research*. New York: Bantam, 1998. Considers the societal implications of the Human Genome Project, particularly in regard to privacy
issues involved in genetic testing and the possible dangers of genetic alteration being used in a new eugenics movement.

Wright, William. *Born that Way: Genes, Behavior, Personality*. New York:
Knopf, 1998. Author believes that genes determine most features of
behavior and personality and offers evidence from research to support his
position, though he states that genetic specifications are not "tyrannical
commands, but rather nudges." He distances himself from the sociopolitical implications of a belief in genetic determinism.

ARTICLES

Beardsley, Tim. "Vital Data." *Scientific American*, vol. 274, March 1996,
pp.100ff. The Human Genome Project is ahead of schedule in its task of
mapping human genes, but many ethical issues remain to be solved,
including those related to commercialization and possible abuses of the
information the project obtains.

Belkin, Lisa. "Splice Einstein and Sammy Glick. Add a Little Magellan." *The
New York Times Magazine*, August 23, 1998, p. 26. Profiles J. Craig Venter,
a molecular biologist leading a privately funded attempt to sequence the
human genome before the Human Genome Project does. Raises questions about his methods and motives.

Collins, Francis, et al. "New Goals for the U.S. Human Genome Project:
1998–2003." *Science*, vol. 282, October 23, 1998, pp. 682ff. Describes
eight new goals for the project, including advancing the target date for
sequencing of the entire human genome from 2005 to 2003.

Elmer-Dewitt, Philip. "The Genetic Revolution." *Time*, vol. 143, January 17,
1994, pp. 46–53. Describes new information coming out of the Human
Genome Project, its potential for treating inherited and perhaps other
diseases, and ethical questions raised by possible misuse of genetic testing.

Greely, Hank. "Mapping the Territory: The Human Genome Diversity
Project Searches for Cures, Not Profits." *Utne Reader*, no. 74,
March–April 1996, pp. 87ff. Unlike the Human Genome Project, which
seeks to map the human genetic code, the Human Genome Diversity
Project attempts to write human genetic history by collecting and analyzing DNA samples from indigenous peoples around the world; any profit
made from the samples is to be shared with the participating population.

Jaroff, Leon. "The Gene Hunt." *Time*, vol. 133, March 20, 1989, pp. 62–67.

Describes the launching of the ambitious $3 billion Human Genome Project, intended to sequence all of the approximately 100,000 genes in a human cell; mentions some of the project's scientific and ethical implications.

Kahn, Patricia. "Genome Diversity Project Tries Again." *Science*, vol. 266, November 4, 1994, pp. 720ff. When it was first proposed in 1991, the Human Genome Diversity Project, intended to survey genetic diversity among the world's indigenous peoples, drew scientific support but also criticism, including accusations of racism and exploitation. In 1994, the project's leader, Luca Cavalli-Sforza, was trying to gather broader-based support.

Lander, Eric S. "The New Genomics: Global Views of Biology." *Science*, vol. 274, October 25, 1996, pp. 536ff. Presents 10 goals to guide the next phase of the massive Human Genome Project and related research.

Marshall, Eliot. "The Genome Program's Conscience." *Science*, vol. 274, October 25, 1996, pp. 488ff. The research program on the ethical, social, and legal implications (ELSI) of the Human Genome Project, launched near the project's beginning by then-director James Watson with 3 percent of the project's budget, is the world's largest bioethics program. Ethicists, geneticists, and genome program officials disagree about its level of success.

"Private Venture to Sequence Human Genome Launched." *Issues in Science and Technology*, vol. 15, Fall 1998, pp. 28–29. The launch of a private venture to sequence the human genome, competing with the government's Human Genome Project, has caused concerns in Congress and among scientists about the quality of the private project's research and possible limitations on access to the information it produces.

Richardson, Sarah. "Forever Young." *Discover*, vol. 20, January 1999, pp. 58, 60. Two laboratories have isolated human stem cells, which give rise to all the tissues in the body. This achievement creates the possibility of growing replacement tissues to order but also raises questions about the ethics of using human embryos for experiments.

Singer, Eleanor, Amy Corning, and Mark Lamias. "The Polls—Trends: Genetic Testing, Engineering, and Therapy: Awareness and Attitudes." *Public Opinion Quarterly*, vol. 62, Winter 1998, pp. 633ff. Provides results of recent opinion polls on questions related to genetic engineering (especially of humans), privacy and possible genetic discrimination, nature vs. nurture, and human cloning.

Specter, Michael. "Decoding Iceland." *New Yorker*, vol. 74, January 18, 1999, pp. 40ff. Describes Icelandic neurologist and Decode Genetics founder Kari Stefansson's plan to catalog genetic information about all of his country's citizens, his project's possible usefulness in analyzing causes of disease, and people's concern about one private company having access to so much information.

Travis, John. "Race to Find Human Stem Cells Ends in Tie." *Science News*, vol. 154, November 7, 1998, p. 293. Two laboratories independently reported isolation of human embryonic stem cells, which are seemingly immortal and can produce any type of cell or tissue in the body.

———. "Another Human Genome Project." *Science News*, vol. 153, May 23, 1998, pp. 334–335. Gene-sequencing expert Craig Venter has teamed with Perkin-Elmer Corporation, the leading maker of automated DNA sequencing machines, in an effort to outrace the government-sponsored Human Genome Project in decoding the human genome.

Wright, Robert. "Achilles' Helix." *New Republic*, vol. 203, July 9 and 16, 1990, pp. 21–31. Describes the scientific and political background of the Human Genome Project and the ethical issues raised by possible uses and misuses of the information it gathers.

INTERNET/WEB DOCUMENTS

Baker, Catherine. *Your Genes, Your Choices: Exploring the Issues Raised by Genetic Research*. American Association for Advancement of Science. Available online. URL: http://ehr.aaas.org/ehr/books/index.html. Downloaded July 1, 1999. Uses fictional case histories to describe the science behind the Human Genome Project and the ethical, legal, and social issues it raises in a manner that young people can understand. Shows how the science of genetics affects choices that people make about their health.

Casey, Denise, and the Human Genome Program, U.S. Department of Energy. *Primer on Molecular Genetics*. Available online. URL: http://www.ornl.gov/TechResources/Human_Genome/publicat/primer/intro.html. Posted 1992. Describes procedures for mapping and sequencing the human genome, different types of maps produced, how the resulting data are stored and interpreted, and impact of the project. Includes glossary.

OTHER MEDIA

"The Human Genome Project." Video. Bethesda, Md.: Public Affairs Department of the National Center for Human Genome Research of the National Institutes of Health, n.d. Promotional video stresses medical and other benefits of the project and compares it to the explorations of the European Renaissance; also considers ethical, legal, and social implications.

DNA "FINGERPRINTING"

Note: Because the most vigorous discussion of DNA profiling took place in its first years of use, between 1989 and 1992, much of the material available on this subject was written before 1995.

BOOKS

Committee on DNA Technology in Forensic Science, National Research Council. *DNA Technology in Forensic Science.* Washington, D.C.: National Academy Press, 1992. Offers recommendations for evaluating DNA testing and applying it in court, including a controversial "ceiling principle" for determining the probability of a match between suspect and crime scene DNA occurring by chance.

————. *An Update: The Evaluation of Forensic DNA Evidence.* Washington, D.C.: National Academy Press, 1996. Revised version of NRC's 1992 report, leaning more toward the prosecution in calculation of statistics and abandoning the "ceiling principle."

Department of Justice, Office of Justice Programs, Bureau of Justice Statistics. *Forensic DNA Analysis: Issues.* Washington, D.C.: U.S. Government Printing Office, 1990. Lays out guidelines for conducting DNA identification testing in criminal cases. These standards are those still in use.

Inman, Keith, and Norah Rudin. *An Introduction to Forensic DNA Analysis.* Grand Rapids, Mich.: CRC Press, 1997. Describes the roots of forensic DNA analysis in genetics, biochemistry, and molecular biology; explains in layperson's terms how the technology works; and considers the technology's advantages and disadvantages from both a scientific and a legal perspective.

Kelly, John F., and Phillip K. Wearne. *Tainting Evidence: Inside the Scandals at the FBI Crime Lab.* New York: Free Press/Simon & Schuster, 1998. Describes how evidence, including DNA evidence, is analyzed at the world's most renowned forensics laboratory and claims that this analysis is marked by both scientific incompetence and personal bias. Has a chapter on the role of DNA testing in the O. J. Simpson trial.

Krawczak, M., and J. Schmidtke. *DNA Fingerprinting.* New York: Springer Verlag, 1999. Describes the technology of DNA testing to identify suspects in criminal cases and to establish family relationships. Considers technical and ethical issues likely to be raised by further development of the technology.

Lampton, Christopher. *DNA Fingerprinting.* Danbury, Conn.: Franklin Watts, 1991. Book for young adults describes the procedures and uses of DNA "fingerprinting" for identification in forensic science.

Levy, Harlan. *And the Blood Cried Out: A Prosecutor's Spellbinding Account of the Power of DNA.* New York: Basic Books, 1996. Author, a prosecutor who is expert in using DNA identification evidence, dissects 14 high-profile crime cases in which such evidence figured and shows how the evidence was evaluated in court in each case.

Mones, Paul. *Stalking Justice.* New York: Pocket Books, 1995. True-crime book tells how police detective Joe Horgas used DNA testing to convict serial rapist-killer Timothy Spencer, who became the first person to be executed because of DNA evidence.

Scheindlin, Gerald. *Genetic Fingerprinting: The Law and Science of DNA.* New York: Routledge, 1996. Author, the judge who presided over the landmark *New York v. Castro* case in 1989, describes how the science of DNA testing is applied to criminal and other legal cases and considers both the technology's power and its pitfalls. Book lists every major DNA case decided by the federal and state courts, as well as state laws related to forensic DNA testing.

Wambaugh, Joseph. *The Blooding.* New York: Perigord, 1989. Describes the first use of DNA "fingerprinting" to identify a criminal, which involved voluntary donation and testing of blood samples from several thousand men in Leicestershire, Britain, in 1987.

ARTICLES

Adler, Jerry, and John McCormick. "The DNA Detectives." *Newsweek,* November 16, 1998, p. 66. Describes current and historical uses of DNA profiling in criminal cases and reviews the possible risks to civil liberties and privacy presented by national DNA databases such as the one established by the FBI.

Aldhouse, Peter. "Geneticists Attack NRC Report as Scientifically Flawed." *Science,* vol. 259, February 5, 1993, pp. 755–756. A coalition of population geneticists and mathematicians questioned some conclusions in the 1992 National Research Council report on DNA profiling, especially its statistical "ceiling principle."

Annas, George J. "Setting Standards for the Use of DNA-typing Results in the Courtroom—The State of the Art." *New England Journal of Medicine,* vol. 326, June 11, 1992, pp. 1641ff. Describes a case in which DNA evidence was challenged on the basis that particular DNA sequences may be more common among certain subsets of the population than in the large groups usually used for computing the probability of a mistaken match.

Balding, Donnelley P. "How Convincing Is DNA Evidence?" *Nature,* vol. 368, 1994, pp. 285–286. Describes criminal cases in which identification by DNA testing has proved doubtful; considers problems of interpretation, including the "prosecutor's fallacy."

Brownlee, Shannon. "Courtroom Genetics." *U. S. News & World Report*, vol. 112, January 27, 1992, pp. 60–61. Describes the battles between experts that often confuse jurors in cases in which suspects are identified by DNA profiling.

———. "Science Takes a Stand." *U.S. News & World Report*, vol. 117, July 11, 1994, pp.29ff. States that DNA "fingerprinting" is likely to play a major role in the O. J. Simpson murder trial and describes points of controversy about the technique that defense lawyers may attack.

Chakraborty, Ranajit, and Kenneth K. Kidd. "The Utility of DNA Typing in Forensic Work." *Science*, vol. 254, December 20, 1991, pp. 1735ff. Rebuttal to the article by Richard Lewontin and Daniel Hartl in the same issue. Maintains that the existence of ethnic subpopulations within large groups such as "Caucasians" does not invalidate the normal method of calculating statistics related to DNA testing.

Cohen, Jack, and Ian Stewart. "Beyond All Reasonable DNA." *Lancet*, vol. 345, June 24, 1995, pp. 1586ff. Considers three problems with interpreting DNA profiling evidence in court: difficulties in determining whether there is a match between suspect and crime scene DNA, contamination, and confusion about statistics.

Devlin, B., Neil Risch, and Kathryn Roeder. "Statistical Evaluation of DNA Fingerprinting: A Critique of the NRC's Report." *Science*, vol. 259, February 5, 1993, pp. 748ff. Claims that the "ceiling principle" for calculating the statistical probability of error in forensic DNA identification, recommended in a 1992 report by the NRC, is arbitrary and is not supported by scientific evidence.

Federal Bureau of Investigation. "Statement of Statistical Standards for DNA Analysis Working Group." *FBI Crime Lab Digest*, vol. 17, March 1990, pp. 53–58. Lists standards for DNA identification testing recommended by the FBI; testing procedures in the United States are still based on these standards.

Franklin-Barbajosa, Cassandra. "DNA Profiling: The New Science of Identity." *National Geographic*, vol. 181, May 1992, pp. 112ff. Describes DNA profiling for disease detection, crime investigation, and other purposes.

Glausiusz, Josie. "Royal D-Loops." *Discover*, vol. 15, January 1994, p. 90. Describes how DNA identification testing was used in 1993 to identify the remains of the last czar of Russia and his family, discovered in a shallow pit in Siberia.

Hawaleshka, Danylo. "A High-Tech Tool for Police." *Maclean's*, vol. 110, March 24, 1997, pp. 56–57. Canada, like the United States, considered establishing a national data bank to help police match DNA from samples found at crime scenes with that of people already convicted of crimes. Article discusses the uses and risks of such a data bank and of DNA identification testing generally.

Kolata, Gina. "U.S. Panel Seeking Restriction on Use of DNA in Courts." *The New York Times*, April 14, 1992, pp. 1, C7. Claims that the 1992 NRC report on DNA testing recommends that DNA evidence not be used in court until testing and related statistics can be made more accurate; report's authors denied this.

Lander, Eric. "DNA Fingerprinting on Trial." *Nature*, vol. 339, 1989, pp. 501–505. Describes Lander's experience as an expert witness in *New York v. Castro*, in which he and other scientists testifying for both the prosecution and the defense banded together to criticize the prosecution's DNA identification evidence; also describes other cases.

Lander, Eric, and Bruce Budowle. "DNA Fingerprinting Dispute Laid to Rest." *Nature*, vol. 371, October 27, 1994, pp. 735–738. Joint article by a longtime critic of DNA testing standards and a leading FBI testing expert, declaring a truce in the "DNA fingerprinting wars" and claiming that the FBI was not involved in the flawed testing attributed to other laboratories.

Lewin, Roger. "DNA Typing on the Witness Stand." *Science*, vol. 244, June 2, 1989, pp. 1033ff. Describes some of the first questioning of forensic DNA testing, including a court case (*New York v. Castro*) in which experts testifying for both the defense and the prosecution agreed that the testing had been done so poorly that its results were invalidated.

Lewis, Ricki. "DNA Fingerprints: Witness for the Prosecution." *Discover*, vol. 9, June 1988, pp. 44–52. Detailed description of the Tommie Lee Andrews rape trials in Orlando, Florida, in late 1987 and early 1998. Andrews became the first American convicted of criminal charges largely on the basis of DNA identification.

———. "Genetics Meets Forensics." *BioScience*, vol. 39, January 1989, pp. 6ff. Describes three DNA testing methods being evaluated at the FBI's forensic research and training laboratory in Quantico, Virginia, and the standard used by courts in deciding whether to accept such evidence. Author says biologists and those in the legal system often have trouble understanding each other concerning technical issues in DNA testing.

Lewontin, Richard C., and Daniel L. Hartl. "Population Genetics in Forensic DNA Typing." *Science*, vol. 254, December 20, 1991, pp. 1745ff. Landmark article questioning the method of determining the probability of a chance match between DNA belonging to two different people, on the grounds that large groups such as "Caucasians" or "Blacks" are made up of subpopulations in which frequency of certain DNA stretches is different from that in the larger population because of increased intermarriage within the subgroup.

Maas, Peter. "Winning Ugly." *Esquire*, vol. 119, April 1993, pp. 83–84. Profiles lawyers Barry Scheck and Peter Neufeld, who have become expert in defending accused people by questioning the results of DNA tests. Describes the weak points in the technology that Scheck and Neufeld attack.

McElfresh, Kevin C., Debbie Vining-Forde, and Ivan Balazs. "DNA-Based Identity Testing in Forensic Science." *BioScience*, vol. 43, March 1993, pp. 149ff. Examines the science behind the technology of forensic DNA identification, the current state of the technology, and its probable future.

Marshall, Eliot. "Academy's About-Face on Forensic DNA." *Science*, vol. 272, May 10, 1996, pp. 803–804. Summarizes recent National Research Council report on DNA identification testing in court cases, in which the NRC removes its controversial 1992 "ceiling principle."

Merz, Beverly. "DNA Fingerprints Come to Court." *Journal of the American Medical Association*, vol. 259, April 15, 1988, pp. 2193–2194. Describes Alec Jeffreys's DNA "fingerprinting" technique and reactions to its first uses in court.

Michaud, Stephen G. "DNA Detectives." *New York Times Magazine*, November 6, 1989, pp. 70–89, 104. Describes the techniques of DNA "fingerprinting," which allow criminals to be identified with unprecedented accuracy.

Moody, Mark D. "DNA Analysis in Forensic Science." *BioScience*, vol. 39, January 1989, pp. 31ff. Describes methods of forensic DNA testing and of calculating the probability of a chance match between the DNA profiles of two people; also mentions nonforensic uses of the technology.

Neufeld, Peter J., and Neville Coleman. "When Science Takes the Witness Stand." *Scientific American*, vol. 262, May 1990, pp. 46–53. Describes the procedure of DNA identification testing in criminal cases and concludes that defense lawyers should question DNA evidence, especially statistics relating to the likelihood of a chance match between suspect and crime scene DNA, more often than they do.

Nowak, Rachel. "Forensic DNA Goes to Court with O. J." *Science*, vol. 265, September 2, 1994, pp. 1352ff. Predicts that forensic DNA identification will be "on trial" along with O. J. Simpson and points out aspects of the technique that defense lawyers are likely to question.

Roberts, Leslie. "DNA Fingerprinting: Academy Reports." *Science*, vol. 256, April 17, 1992, pp. 300–301. Summarizes a long-awaited report on forensic DNA identification by the NRC of the National Academy of Sciences. The report attempts to resolve conflicts concerning the validity of statistics that determine the probability of a chance match between suspect and crime scene DNA.

———. "Fight Erupts Over DNA Fingerprinting." *Science*, vol. 254, December 20, 1991, pp. 1721–1722. Summarizes the disagreement between population geneticists over the validity of statistics used to determine the probability of a false match in DNA profiling, laid out in opposing articles elsewhere in the same issue of the magazine. Also describes alleged attempts to keep one of the articles from being published.

———. "Science in Court: A Culture Clash." *Science*, vol. 257, August 7, 1992, pp. 732ff. Describes the "culture shock" of scientists testifying as expert witnesses about DNA profiling in the *United States v. Yee* case, where they found their honesty and motives impugned by opposing lawyers.

Roeder, K. "DNA Fingerprinting: A Review of the Controversy." *Statistics Science*, 1994, no. 9, pp 222–278. Discusses the three main concerns about DNA evidence in court: technical error in testing, genetic differences in human subpopulations, and misunderstanding of the meaning of statistical claims made by expert witnesses.

Taylor, Charles. "Survey of Population Geneticists Concerning Methods for Calculating Matches in Forensic Applications of VNTR." *Loci*, July 9, 1991. States that of 30 population geneticists surveyed, 11 supported the FBI's method for calculating the probability of random matches in DNA testing; 19 did not.

Thompson, William C. "Accepting Lower Standards: The National Research Council's Second Report on Forensic DNA Evidence." *Jurimetrics*, vol. 37, 1997, pp. 405–424. Compares the 1992 and 1996 NRC reports on forensic DNA testing and concludes that the revised report is inferior to the earlier one.

Vogel, Shawna. "The Case of the Unraveling DNA." *Discover*, vol. 11, January 1990, pp. 46–47. Describes the murder trial of José Castro, in which experts testifying for both the defense and the prosecution took the unusual step of meeting outside the courtroom and concluding that the DNA evidence that the prosecution planned to present in the trial was seriously flawed.

Wall, W. J. "Whose DNA Is It Anyway?" *New Statesman and Society*, vol. 8, December 8, 1995, pp. 20–21. Describes a new technique of DNA profiling, the Short Tandem Repeat (STR) method, which combines features of the two most commonly used older methods; maintains that although the technique is an improvement, it does not solve all problems associated with previous methods.

INTERNET/WEB DOCUMENTS

Harvey, James Virgil. "Acquit or Convict: Should Profile DNA Be Conclusive?" Available online. URL: http://louweb1.mis.net/sdk/winner.htm. Posted 1997. This essay, which won the Sigma Delta Kappa Law Foundation Excellence in Legal Writing Competition, discusses the techniques of DNA identification profiling, the pretrial hearing that determines whether DNA evidence will be admitted in a particular case, and problems on the basis of which DNA evidence can be challenged in court.

Kaye, D. H. "DNA Identification: Some Lingering and Emerging Evidentiary Issues." Available online. URL: http://www.law.asu.edu/personal/kaye/pubs/dna/promega7.htm. Posted December 1997. This paper, presented at the Seventh International Symposium on Human Identification in December 1997, considers problems with DNA identification evidence, including sources of error in interpreting test results and calculating match statistics.

Sylvester, John T., and John H. Stafford. "Judicial Acceptance of DNA Profiling." Available online. URL: http://www.phreak.org/archives/The_Hacker_Chronicles_II/lawnt/judna.txt. Downloaded July 1, 1999. The authors, special agents of the DNA Task Force of the FBI's Legal Counsel Division, discuss key court cases in which the validity of DNA profiling evidence has been questioned. They conclude that most courts have accepted and will continue to accept such evidence in spite of problems that have occurred.

Trebach, Susan, et al. "Forensic Science." Available online. URL: http://whyfiles.news.wisc.edu/014forensic/index.html. Posted May 9, 1996. Part of the Why Files, a project funded by the National Science Foundation, this online article provides a lively introduction to forensic science techniques, including DNA fingerprinting.

GENETIC HEALTH TESTING AND DISCRIMINATION

Note: Articles that describe eugenics primarily as a form of genetic discrimination appear in this section. Articles that focus on the possibility of a "new eugenics" involving human gene alteration appear in the "Gene Therapy and Human Gene Alteration" section.

BOOKS

Andrews, Lori B., et al. *Assessing Genetic Risks: Implications for Health and Social Policy*. Washington, D.C.: National Academy Press, 1994. Points out the importance of educating the public about the ethical, legal, and social implications of new discoveries in genetics and genetic testing, including the risk of discrimination in insurance and employment; makes recommendations for research and policy.

Dowbiggin, Ian Robert. *Keeping America Sane: Psychiatry and Eugenics in the United States and Canada, 1880–1940*. Ithaca, N.Y.: Cornell University Press, 1997. Describes how feelings of vulnerability and desire to keep up with the latest science led American psychiatrists in the first half of this century to accept the dubious doctrine of eugenics and conclude that forced sterilization of the mentally ill was justified.

Harper, Peter S., and Angus J. Clarke. *Genetics, Society, and Clinical Practice*. Herndon, Va.: Bios Scientific Publications Ltd., 1997. Collection of essays defines genetic testing and discusses the issues it raises, including testing of children and fetuses, effect of genetic test results on insurance, challenges to genetic privacy, and abuses of human genetic information.

Holtzman, Neil A., and Michael S. Watson, eds. *Promoting Safe and Effective Genetic Testing in the United States: Final Report*. Baltimore: Johns Hopkins University Press, 1998. Final report of the Task Force on Genetic Testing of the National Human Genome Research Institute's Ethical, Legal, and Social Implications (ELSI) Working Group describes methods for ensuring accuracy and privacy of genetic testing. Topics covered include informed consent, prenatal and carrier testing, testing of children, confidentiality, and discrimination.

Human Genetics Advisory Commission. *The Implications of Genetic Testing for Insurance*. London: Office of Science and Technology, 1997. Report by a British government commission warns that growing use of genetic tests may lead to increased discrimination in insurance; makes recommendations for legislation and regulation to prevent this.

Kevles, Daniel J. *In the Name of Eugenics: Genetics and the Uses of Human Heredity*. Cambridge: Harvard University Press, 1995. Traces the rise of the eugenics movement from its beginnings in the late 1800s to its more subtle modern incarnation, discrimination on the basis of genetic screening.

Lynn, Richard. *Dysgenics*. Westport, Conn.: Praeger, 1996. Author provides a history of the eugenics movement in the late 19th and early 20th centuries, explaining why belief in eugenics came to be widely accepted and, later, just as widely dismissed. He seeks to rehabilitate the argument that genetic deterioration is occurring in the human species.

Pernick, Martin S. *The Black Stork: Eugenics and the Death of "Defective" Babies in American Medicine and Motion Pictures Since 1915*. New York: Oxford University Press, 1996. Beginning with a well-publicized case in the late 1910s in which a doctor let six babies he considered "defective" die, this book traces the history of the eugenics movement and links it to euthanasia and to race, class, gender, and ethnic hatreds, showing the interaction among medical concepts, cultural values, and media portrayals. Book concentrates on the early 20th century, but the issues it discusses have implications for today.

Rothstein Mark A., ed. *Genetic Secrets: Protecting Privacy and Confidentiality in the Genetic Era*. New Haven: Yale University Press, 1997. Experts survey legal and ethical issues raised by the growing ability to obtain detailed genetic information that can predict future health, including implications for the doctor-patient relationship, medical privacy, discrimination, availability and cost of insurance, and conduct of medical research.

Smith, J. David. *The Eugenic Assault on America: Scenes in Red, White, and Black*. Fairfax, Va.: George Mason University Press, 1993. Chilling account of the rise in the 1920s and 1930s of state eugenics laws ordering forced sterilization of the "feebleminded," insane, criminal, or others deemed unfit. Author explores the racial aspects of the eugenics movement and shows that the challenges to civil liberties it presented are far from dead.

Teichler-Zallen, Doris. *Does It Run in the Family? A Consumer's Guide to DNA Testing for Genetic Disorders*. Rutgers, N.J.: Rutgers University Press, 1997. A geneticist and science policy expert explains how genetic disorders are inherited, which ones can be tested for, how the tests work and what they can and cannot show, and how families can make wise choices about being tested; includes discussion of genetic testing in relation to health care, government and law, and insurance.

Wingerson, Lois. *Unnatural Selection*. New York: Bantam, 1998. Focuses on potential dangers posed by the information gathered by the Human Genome Project, including threats to privacy and genetic diversity, medical dilemmas, and possible genetic discrimination.

ARTICLES

Allen, Garland E. "Science Misapplied: The Eugenics Age Revisited." *Technology Review*, vol. 99, August–September 1996, pp. 22ff. Detailed history of the eugenics movement in the late 19th and early 20th centuries stresses its social and economic roots and expresses fears that the combination of genetic determinism and stress on cost effectiveness could produce a similar movement in the near future.

American Council of Life Insurance and the Health Insurance Association of America. "Report of the ACLI-HIAA Task Force on Genetic Testing." 1995. Maintains that only a small percentage of the population is refused life insurance or charged higher than normal rates because of genetic test results.

American Society for Human Genetics and American College of Medical Genetics. "Points to Consider: Ethical, Legal, and Psychosocial Implications of Genetic Testing in Children and Adolescents." *American*

Journal of Human Genetics, no. 57, 1995, pp. 1233–1241. Considers questions of consent, discrimination, and other ethical, legal, and social aspects of doing genetic tests on minors.

Association of British Insurers. "Information Sheet: Life Insurance and Genetics." London: Association of British Insurers, 1997. Describes association's new policy on use of information from genetic testing.

Baird, P. A. "Identifying People's Genes: Ethical Aspects of DNA Sampling in Populations." *Perspectives in Biological Medicine*, no. 38, 1995, pp. 159–166. Considers ethical aspects of screening healthy populations for carriers of inherited diseases.

Basu, Janet. "Genetic Roulette." *Stanford Today*, November–December 1996, pp. 38–43. Describes some of the ethical, legal, and social implications of information gathered by the Human Genome Project. Focuses on misuse of information from genetic tests and resultant discrimination, which has already occurred.

Beardsley, Tim. "China Syndrome." *Scientific American*, vol. 276, March 1997, pp. 33–34. Describes a current Chinese law requiring couples with a history of genetic defects in their families to undergo sterilization or long-term contraception. Claims that multinational biotechnology companies plan to use the Chinese gene pool for drug research.

Cowley, Geoffrey. "Flunk the Gene Test and Lose Your Insurance." *Newsweek*, vol. 128, December 23, 1996, pp. 48–50. Genetic testing can provide useful information about health risks, but author maintains that it has already led to discrimination in insurance, employment, and marriage.

Dikotter, Frank. "'The Legislation Imposes Decisions.'" *UNESCO Courier*, September 1999, p. 31. Describes classist and racist implications of China's 1995 Maternal and Infant Health Law and questions whether the "informed consent" required by the law has much meaning. Includes excerpts from the law.

———. "Race Culture: Recent Perspectives on the History of Eugenics." *American Historical Review*, vol. 103, April 1998, pp. 467ff. Points out that practices related to eugenics have existed not only in the well-studied countries of Great Britain, the United States, and Germany but also in lesser known areas, such as Latin America and France.

Dreyfus, Rochelle Cooper, and Dorothy Nelkin. "The Jurisprudence of Genetics." *Vanderbilt Law Review*, vol. 45, 1992. Survey of legal issues involving genetics, including privacy and misuse of genetic information.

Gaulding, J. "Race, Sex and Genetic Discrimination in Insurance: What's Fair?" *Cornell Law Review*, no. 80, 1995, pp. 1645–1646. Considers legal and ethical aspects of differential risk calculation in insurance based on race, gender, or genetic profile.

Annotated Bibliography

Geller, Lisa, et al. "Individual, Family, and Societal Dimensions of Genetic Discrimination: A Case Study Analysis." *Science and Engineering Ethics*, vol. 2, 1996. This Harvard University survey suggests that discrimination on the basis of genetic testing is relatively widespread in the United States and occurs in several types of organizations.

Goode, Stephen. "Marines Stand Ground Against DNA Testing." *Insight on the News*, vol. 12, February 19, 1996, p. 38. Marines John C. Mayfield III and Joseph Vlacovsky faced courts martial for refusing to give samples of their DNA to a Department of Defense database, claiming that being required to do so violated their privacy. (They were later convicted.)

Hallowell, Christopher. "Playing the Odds." *Time*, vol. 153, January 11, 1999, p. 60. Insurance organizations say people's fears of discrimination on the basis of genetic tests are exaggerated, and legislation is being designed to prevent such discrimination, but fears persist that insurance carriers and HMOs will use genetic information to weed out high-risk candidates as a way of keeping costs down and profits up.

Harper, Peter S. "Genetic Testing, Common Diseases, and Health Service Provision." *Lancet*, vol. 346, December 23, 1995, pp. 1645–1646. Recommends cautious use of genetic screening tests, especially those whose reliability has not been proven, because indiscriminate use of such tests could produce unnecessary expense, misunderstanding, and possible psychological or social damage, including discrimination in health insurance.

Hawkins, Dana. "Court Declares Right to Genetic Privacy." *U.S. News & World Report*, vol. 124, February 16, 1998, p. 4. Reports that in a case involving Lawrence Berkeley Laboratory in California, a U.S. Court of Appeals has ruled that performing genetic testing without consent violates Fourth Amendment privacy rights and that testing African Americans and Hispanics to a greater extent than Caucasian also violates the Civil Rights Act of 1964.

———. "Dangerous Legacies: New Gene Tests Provide Fresh Grounds for Discrimination." *U.S. News & World Report*, vol. 123, November 10, 1997, pp. 99–100. Reports that many women with family histories of breast and ovarian cancer are afraid to be tested for genes that carry increased risk of these diseases because they fear discrimination by employers and insurers. Women's health advocates want federal legislation to protect people against such discrimination.

Helmuth, L. "Disability Law May Cover Gene Flaws." *Science News*, vol. 155, February 27, 1999, p. 134. The 1998 Supreme Court ruling in *Bragdon v. Abbott* suggests that the Americans with Disabilities Act can be called on in a legal strategy to protect healthy people who are genetically predisposed to an illness from discrimination.

"Here, of All Places: Nordic Eugenics." *Economist*, vol. 344, August 30, 1997, pp. 36–37. Points out that not only the United States and Nazi Germany but the Scandinavian countries—Denmark, Finland, Norway, and Sweden—had eugenics laws that permitted forcible sterilization of those deemed mentally defective. The laws remained on the books from the 1930s to the 1970s.

Hilgers, Laura. "Should You Take a Genetic Test? Not Before You Read This." *Glamour*, vol. 94, October 1996, pp. 272ff. Warns women considering being tested for genes that predispose to breast and ovarian cancer that they may face discrimination by insurers and employers based on the results.

Hudson, Kathy L., et al. "Genetic Discrimination and Health Insurance: An Urgent Need for Reform." *Science*, vol. 270, October 20, 1995, pp. 391ff. Claims that discrimination in health insurance on the basis of genetic information can have devastating consequences for those discriminated against and can also slow genetic research because it makes people unwilling to undergo genetic testing. Authors recommend state and federal legislation to bar such discrimination.

Jones, Valerie A. "In the Same Boat." *Journal of the American Medical Association*, vol. 280, November 4, 1998, p. 1537. Warns against excessive use of genetic screening because such screening can distress patients, increase health care costs, and invite discrimination by insurers and others.

Kevles, Bettyann H., and Daniel J. Kevles. "Scapegoat Biology." *Discover*, vol. 18, October 1997, pp. 58–62. Considers the likelihood that attempts to identify genetic causes of violent behavior will be used as excuses for discrimination and for ignoring social causes of such behavior.

Kirby, Michael. "Genetic Testing and Discrimination." *UNESCO Courier*, May 1998, pp. 29ff. States that organizations and governments need to develop policies to protect people from discrimination by insurance companies or others, based on the information revealed by genetic testing.

Kodish, Eric, et al. "Genetic Testing for Cancer Risk: How to Reconcile the Conflicts." *Journal of the American Medical Association*, vol. 279, January 21, 1998, pp. 179–181. Compares three conflicting policy directives regarding the risks (including discrimination) and benefits of genetic screening, using a recently developed test for a gene that predisposes to breast cancer as an example.

Korn, David. "Dangerous Intersections." *Issues in Science and Technology*, vol. 13, Fall 1996, pp. 55ff. Argues that while researchers must take the potential "psychosocial risk" of genetic tests seriously, proposed legislation aimed at protecting privacy may hamper promising research.

Lapham, E. Virginia, Chahira Kozma, and Joan O. Weiss. "Genetic Discrimination: Perspectives of Consumers." *Science*, vol. 274, October 25, 1996, pp. 621ff. A study of 332 members of support groups for inherited disorders showed widespread feelings that respondents or affected family members had been discriminated against in insurance or employment or fears that this would happen. More research is needed to discover how accurate these perceptions are.

Mehlman, M. J., et al. "The Need for Anonymous Genetic Counseling and Testing." *American Journal of Human Genetics*, no. 58, 1996, pp. 393–397. Maintains that anonymity is necessary to protect patients' privacy and prevent genetic discrimination.

Natowitcz, Marvin R., et al. "Genetic Discrimination and the Law." *American Journal of Genetics*, no. 50, 1992. Surveys the laws bearing on discriminatory use of genetic information.

Nemeth, Mary. "'Nobody Has the Right to Play God.'" *Maclean's*, vol. 108, June 26, 1995, p.17. Describes the case of Leilani Muir, a woman who was incorrectly diagnosed as a mental defective and forcibly sterilized under Alberta's former eugenics law; she is suing the government.

Normile, Dennis. "Geneticists Debate Eugenics and China's Infant Health Law." *Science*, vol. 281, August 21, 1998, pp. 1118–1119. Chinese scientists told a meeting of geneticists that a poor translation made China's recent "eugenics" law seem more draconian than it is. They said that the law is seldom enforced because of a lack of agreement on what inherited conditions it should cover and a lack of facilities for genetic testing.

Pokorski, R. J. "Insurance Underwriting in the Genetic Era." *American Journal of Human Genetics*, no. 60, 1997, pp. 205–216. Describes how genetic testing will affect calculation of risks for health and life insurance.

Ponder, Bruce. "Genetic Testing for Cancer Risk." *Science*, vol. 278, November 7, 1997, pp. 1050ff. Describes controversies surrounding tests for cancer-related genes, including possible loss of privacy and discrimination in insurance and employment; calls for careful evaluation of the costs and benefits of such tests.

Rothenberg, Karen, et al. "Genetic Information and the Workplace: Legislative Approaches and Policy Changes." *Science*, vol. 275, March 21, 1997, pp. 1755–1757. Presents joint recommendations of the National Action Plan for Breast Cancer's Hereditary Susceptibility Working Group and the National Human Genome Research Institute's Ethical, Legal and Social Implications (ELSI) Working Group for laws and regulations to limit employment discrimination based on genetic information.

"Roundtable: The Politics of Genetic Testing." *Issues in Science and Technology*, vol. 13, fall 1996, pp. 48ff. Discussion held during a conference on "The Genetics Revolution" in March 1996 considers political and ethical issues raised by genetic testing, including those of privacy and potential discrimination.

Shulman, Seth. "Preventing Genetic Discrimination." *Technology Review*, vol. 98, July 1995, pp. 16ff. Describes a California law, passed in 1994, that prohibits employers and insurers from discriminating against healthy people genetically predisposed to rare diseases; physicians are also barred from releasing genetic test information to insurance companies or employers.

Thomson, B. "Time for Reassessment of Use of All Medical Information by UK Insurers." *Lancet*, October 10, 1998, p. 1216. Argues that attempts to prohibit use of genetic-based information by insurers is likely to fail because of the difficulty of defining what information is really genetic-based; suggests looking instead at the fundamental question of what is a private responsibility and insurable and what must be covered by government.

Townsend, Kathleen Kennedy. "The Double-Edged Helix." *Washington Monthly*, vol. 29, November 1997, pp. 36–37. Maintains that genetic discrimination can keep people from getting disability insurance and that a national health program is necessary to prevent such discrimination from occurring.

Yesley, Michael S. "Protecting Genetic Difference." *Berkeley Technology Law Journal*, vol. 13, 1998, pp. 653–665. Discusses recent legal changes related to protection of genetic privacy and prohibition of genetic discrimination.

INTERNET/WEB DOCUMENTS

National Action Plan for Breast Cancer, Hereditary Susceptibility Working Group. "*Bragdon v. Abbott*: Implications for Asymptomatic Genetic Conditions." Available online. URL: http://www.napbc.org/napbc/heredita1.htm. Posted February 19, 1999. Summary of a workshop in which participants consider whether this key case, in which the Supreme Court ruled that a woman who was infected with HIV but had no symptoms of illness could be classified as disabled under the Americans with Disabilities Act (ADA), means that healthy people shown by genetic tests to be susceptible to disease might also be covered by the ADA.

National Action Plan for Breast Cancer and NIH/DOE ELSI Working Group. *Recommendations on Genetic Information and the Work-place.* Available online. URL: http://www.napbc.org/napbc/recommen.htm. Posted 1997. Recommends legislation that would bar employers from requesting or using genetic information about prospective or existing employees unless it could be shown to be job related and consistent with business necessity.

OTHER MEDIA

Human Genome Education Model (HuGEM) Project. *An Overview of the Human Genome Project and Its Ethical, Legal, and Social Issues.* Video. Washington, D.C.: Georgetown University Medical Center, 1995. Includes discussion of the risks of discrimination against people with genetic defects in health insurance and possible devastating effects of such discrimination.

National Action Plan on Breast Cancer. *Genetic Testing for Breast Cancer Risk: It's Your Choice.* Video. Washington, D.C.: National Action Plan on Breast Cancer. This video, narrated by ABC News correspondent Cokie Roberts, attempts to answer women's questions about genetic testing and provide a balanced view of what test results mean for them and their families. Video comes with a fact sheet and brochure.

GENE THERAPY AND HUMAN GENE ALTERATION

Note: Books and articles that consider eugenics primarily in connection with future human gene alteration appear in this section. Books and articles that consider eugenics primarily as a form of genetic discrimination appear in the "Genetic Health Testing and Discrimination" section.

BOOKS

Avise, John C. *The Genetic Gods: Evolution and Belief in Human Affairs.* Cambridge, Mass.: Harvard University Press, 1998. Claims that humans are using their newfound ability to change genes to reshape themselves and their environment both intellectually and technologically; considers how these changes might affect the human condition and change religious beliefs.

Bova, Ben. *Immortality: How Science Is Extending Your Lifespan, and Changing the World.* New York: Avon, 1998. Author, a well-known science fiction writer, describes how present and near-future biomedical advances, especially in gene therapy, are likely to conquer aging and extend human life as well as curing many diseases and making organ transplants unnecessary. Book gives little analysis of the possible sociopolitical effects of greatly increased lifespans.

Clark, William R. *The New Healers: The Promise and Problems of Molecular Medicine in the Twenty-First Century.* New York: Oxford University Press, 1997. Examines the development of gene therapy, problems still to be solved, and the technique's promise in treating as many as 4,000 inherited human diseases as well as other diseases with a genetic component, such as cancer and AIDS.

Cole-Turner, Ronald, and Brent Waters. *Pastoral Genetics: Theology and Care at the Beginning of Life.* Louisville, Ky.: Westminster John Knox, 1996. Considers genetics, especially genetic decisions and treatments affecting fetuses and newborns, as a pastoral issue and discusses how to connect God with genetic and reproductive processes.

Gosden, Roger. *Designing Babies: The Brave New World of Reproductive Technology.* New York: W. H. Freeman, 1999. Using literary and philosophical as well as medical anecdotes, an expert in the new reproductive medicine puts advances in the field in a cultural context. He considers the social and ethical dilemmas that the technology presents but ultimately takes an optimistic view of a future in which parents will be able to choose or alter their offspring's genes at will.

Heyd, David. *Genetics: Moral Issues in the Creation of People.* Berkeley: University of California Press, 1992. Examines the philosophical significance of the growing belief in the importance of genes in shaping human identity and the ability to alter human genes before birth; discusses why traditional ethical theories have failed to deal with these issues and argues for creation of a new "genethics."

Kimbrell, Andrew. *The Human Body Shop: The Engineering and Marketing of Life.* San Francisco: HarperSanFrancisco, 1993. Kimbrell, Jeremy Rifkin's lawyer, provides a cautionary view of tissue and organ transplants, new reproductive technologies, and alteration of human genes.

Lyon, Jeff, and Peter Gorner. *Altered Fates: Gene Therapy and the Retooling of Human Life.* New York: W. W. Norton, 1995. Describes the history and possible future of gene therapy, bringing out both the positive views of the technology held by the scientists who developed it and the more negative implications that they chose not to consider.

McGee, Glenn. *The Perfect Baby: A Pragmatic Approach to Genetics.* Lanham, Md.: Rowman & Littlefield, 1997. Author bases his discussion of ethical issues raised by new reproductive technologies (such as those related to identity, illness, enhancement, and perfection) on the pragmatism of American philosophers William James and Thomas Dewey.

Maranto, Gina. *Quest for Perfection: The Drive to Breed Better Human Beings.* New York: Scribner, 1996. Considers the moral implications of recent reproductive advances and the dangers of trying to alter the human evolutionary path and create "perfect" human beings; argues for the sanctity of life.

Paul, Diane B. *Controlling Human Heredity: 1865 to the Present.* Buffalo, N.Y.: Prometheus Books, 1995. Offers a history of eugenics, including how it came to be so widely appealing and why it fell into disrepute, and discusses its implications for modern genetic medicine.

Peters, Ted. *For the Love of Children: Genetic Technology and the Future of the Family.* Louisville, Ky.: Westminster John Knox, 1996. Discusses disturbing religious and ethical issues raised by new reproductive technologies, including human gene alteration, but ultimately concludes that these technologies can help to create a society that fosters the dignity and well-being of each child regardless of biological relationships.

Reich, Warren Thomas, ed. *Bioethics: Sex, Genetics, and Human Reproduction.* New York: Macmillan, 1997. Extracts articles from Macmillan's five-volume *Encyclopedia of Bioethics* that pertain to the moral, medical, legal, and social issues surrounding sex and reproduction, including the possibility of genetic selection or alteration in offspring.

Resnik, David, Holly B. Steinkraus, and Pamela J. Langer. *Human Germline Gene Therapy: Scientific, Moral, and Political Issues.* Georgetown, Tex.: R. G. Landes Co., 1999. Describes present gene therapy and the additional technology that would be required to extend therapy to the germ line, where it would affect not only an individual but all of that person's descendants. Authors offer a sober evaluation of the promises and dangers of this near-future technology from ethical, political, and scientific standpoints and make suggestions for international policy to regulate it.

Roberts, Melinda A. *Child versus Childmaker: Future Persons and Present Duties in Ethics and the Law.* Lanham, Md.: Rowman & Littlefield, 1998. Considers philosophical and ethical aspects of human cloning and other new or likely reproductive technologies, including responsibilities to future generations; defends a "a person-affecting solution" to conflicts of interest between "childmakers" and the children they aim to create.

Rothblatt, Martine Aliana. *Unzipped Genes: Taking Charge of Baby-Making in the New Millennium.* Philadelphia: Temple University Press, 1997. Author believes that society will learn to manage the many choices made available by new genetic and reproductive technologies, including the temptation toward personal and social eugenics, and will become able both to minimize disease and to improve humanity's genetic endowment.

Rosenberg, Steven A., and John M. Barry. *The Transformed Cell.* New York: Putnam, 1992. Describes Rosenberg's attempts to boost the immune system's ability to fight cancer, including his use in 1988 of inserted genes (which had no medical effect in themselves) as markers for immune cells altered in the laboratory and reinjected into patients—the first insertion of engineered genes into humans.

Silver, Lee M. *Remaking Eden: Cloning and Beyond in a Brave New World.* New York: Avon, 1997. Optimistic book on human reproductive technologies, including possible cloning and altering genes of the unborn; mainly descriptive, with fictional/predictive scenarios.

Steinberg, Deborah Lynn. *Bodies in Glass: Genetics, Eugenics, Embryo Ethics.* Manchester, England: Manchester University Press, 1997. Expresses concern that increased ability to detect "defective" genes in screening tests and to alter them or select embryos that do not have them will lead to excessive control of human reproduction by authorities.

Thompson, Larry. *Correcting the Code.* New York: Simon & Schuster, 1994. Tells the exciting story of the first uses of human gene therapy and the experiments and political wrangling that led up to them.

Walters, Leroy, and Julie Gage Palmer. *The Ethics of Human Gene Therapy.* New York: Oxford University Press, 1996. Provides background information on human genetics; describes somatic (body cell) and germ-line (sex cell) gene therapy; and considers the ethical issues they raise.

ARTICLES

Anderson, W. French. "Gene Therapy." *Scientific American*, vol. 273, September 1995, pp. 124–128. Presents the scientific background for Anderson's treatment of four-year-old Ashanthi deSilva in 1990, the first use of inserted genes to treat a human illness; also describes the treatment's results.

———. "Human Gene Therapy." *Science*, vol. 256, May 8, 1992, pp. 808–813. Technical description of the first insertions of genes into humans, including Steven Rosenberg's use of medically inactive marker genes in immune cells given as a cancer treatment in 1988 and Anderson's own use of inserted genes to treat a girl with an inherited enzyme deficiency two years later; also covers future prospects for human gene therapy.

Billings, Paul R., Ruth Hubbard, and Stuart A. Newman. "Human Germline Gene Modification: A Dissent." *Lancet*, vol. 353, May 29, 1999, p. 1873. Authors say that germ-line genes should never be altered, even to prevent disease, because doing so may produce unpredictable effects that affect an individual's descendants.

Blaese, R. Michael. "Gene Therapy for Cancer." *Scientific American*, vol. 276, June 1997, pp. 111–115. Gene therapy for cancer focuses on immunotherapy or "cancer vaccines," which increase the power of the immune system to recognize and destroy tumor cells.

Bylinsky, Gene. "Cell Suicide: The Birth of a Mega-Market." *Fortune*, vol. 131, May 15, 1995, pp. 75ff. The next wave of medical biotechnology may focus on drugs or inserted genes that persuade certain cells (such as cancer cells or cells infected with HIV) to kill themselves, a natural process called apoptosis.

Couzen, Jennifer. "RAC Confronts in Utero Gene Therapy." *Science*, vol.

282, October 2, 1998, p. 27. The NIH's Recombinant DNA Advisory Committee has started debating whether to permit gene therapy in fetuses before birth, which is liable to be attempted within the next few years. Such treatment has potential new risks and ethical implications, including the possibility of affecting germ-line genes.

Crystal, Ronald G. "Transfer of Genes to Humans: Early Lessons and Obstacles to Success." *Science*, vol. 270, October 20, 1995, pp. 404–409. Clinical trials and experimental treatments have shown that transferring genes into humans can be done, can have significant medical effects, and is usually safe, but many technical problems must still be solved before it becomes a common form of therapy.

Felgner, Philip L. "Nonviral Strategies for Gene Therapy." *Scientific American*, vol. 276, June 1997, pp. 102–106. Using viruses as vectors to deliver genes to cells can result in immune reactions that destroy the vectors. More effective gene therapy strategies now being developed include injecting "naked" DNA and coating the DNA with fatty substances that do not provoke an immune response.

Freundlich, Naomi. "Finding a Cure in DNA?" *Business Week*, no. 3517, March 10, 1997, pp. 90–92. A new branch of biotechnology called genomics attempts to go beyond sequencing human genes to understanding how those genes cause diseases. Companies working in this field could eventually cure important diseases and generate immense profits.

Friedmann, Theodore. "Overcoming the Obstacles to Gene Therapy." *Scientific American*, vol. 276, June 1997, pp. 96–101. The greatest challenge for gene therapy is designing ways to get genes into cells safely and effectively. Problems with viruses, the most commonly used vectors, include the fact that they insert genes into the target genome at random locations.

Gordon, Jon W. "Genetic Enhancement in Humans." *Science*, vol. 283, March 26, 1999, pp. 2023. Genetic enhancement of humans (that is, improvements in normal people's genes, as opposed to gene therapy) is not likely to be practical in the near future and, in any case, will probably never become widespread, author maintains. He is opposed to human gene enhancement but feels that laws banning it are unlikely to be effective.

Gregory, Tanya, and Nelson A. Wivel. "Clinical Applications of Molecular Medicine." *Patient Care*, November 15, 1998, pp. 86ff. Describes possible use of gene therapy in treating not only inherited deficiency diseases but such conditions as cancer, AIDS, and cardiovascular disease.

Henig, Robin Marantz. "Dr. Anderson's Gene Machine." *New York Times Magazine*, March 31, 1991, pp. 31–35, 50. Dramatic account of W. French Anderson's use of gene therapy to treat four-year-old Ashanthi deSilva, who suffered from an inherited disease caused by a missing enzyme, in September 1990—the first use of inserted genes to treat human illness.

———. "Tempting Fates." *Discover*, vol. 19, May 1998, pp. 52–64. If parents could have their children's genes changed in order to prevent or cure disease, or even to give the offspring an edge in job competition, they might feel it their duty to do so—but changing germ-line genes, which would affect all future descendants, could have unexpected, disastrous consequences.

Ho, Dora Y., and Robert M. Sapolsky. "Gene Therapy for the Nervous System." *Scientific American*, vol. 276, June 1997, pp. 116–120. New discoveries about the way neurons die and are replaced and new techniques for delivering genes to nerve tissue could create novel ways to treat progressive neurological disorders such as Alzheimer's and Parkinson's diseases.

Jaroff, Leon. "Keys to the Kingdom." *Time*, vol. 148, Fall 1996, pp. 24–29. Offers optimistic predictions about the future benefits of gene therapy; also covers genetic testing.

Kass, Leon R. "The Moral Meaning of Genetic Technology." *Commentary*, vol. 108, September 1999, pp. 32ff. Provides ethical (but nonreligious) reasons for being suspicious of altering human genes. Claims that doing so may change the basic nature and definition of humanity.

Lemonick, Michael D., Alice Park, and Clare Thompson. "On the Horizon." *Time*, vol. 153, January 11, 1999, p. 89. Describes new forms of gene therapy and other medical treatments arising from biotechnology that are likely to be developed in the next decade or so, including new kinds of vaccines, new sources of transplantable tissue, and techniques to slow or reverse aging.

Miller, Henry I. "Gene Therapy for Enhancement." *Lancet*, vol. 344, July 30, 1994, pp. 316–317. Considers the ethics of using gene therapy to improve characteristics of normal people, including the fact that such alterations would probably be available only to the wealthy.

Mirsky, Steve, and John Rennie. "What Cloning Means for Gene Therapy." *Scientific American*, vol. 276, June 1997, pp. 122–123. Cloning could be combined with gene therapy to change human germ-line genes, thus removing fatal or debilitating inherited disorders permanently from the human race. Couples carrying a genetic illness could also have the condition corrected by applying gene therapy to an embryo.

Office of Recombinant DNA Activities. "Human Gene Therapy Protocols." National Institutes of Health, July 13, 1998. Lists procedures that federally funded researchers on human gene therapy are expected to follow.

Pool, Robert. "Saviors." *Discover*, vol. 19, May 1998, pp. 53–57. Genetic engineering and biotechnology may eventually solve the tissue and organ transplant shortage by allowing transplants to be made from animals carrying human genes or from tissues grown in the laboratory.

Raloff, Janet. "New Gene Therapy Fights Frailty." *Science News*, vol. 154, December 19 and 26, 1998, p. 388. Inserting a gene that codes for a growth hormone into the muscles of mice offset the muscle wasting that occurs in old age, spurring hope that such therapy might eventually be used to help older humans or those with heart muscle damaged by a heart attack.

"The Real Biotech Revolution." *Fortune*, vol. 135, March 31, 1997, pp. 54–55. Genomics, a new branch of biotechnology, promises to extend gene therapy from mere treatment of disease to permanent cure or prevention.

Regelado, Antonio. "The Troubled Hunt for the Ultimate Cell." *Technology Review*, vol. 101, July–August 1998, pp. 34ff. Describes history of the hunt for human embryonic stem cells, which theoretically could be used to generate any type of tissue in the body and thus provide tissue replacements as needed. Includes discussion of support of this research by biotechnology companies and opposition to it by groups who oppose use of human embryos in medical research.

Rifkin, Jeremy. "The Sociology of the Gene." *Phi Delta Kappan*, vol. 79, May 1998, pp. 648ff. Advances in biotechnology and gene therapy make some people see the "perfection" of human nature through genetic engineering as a desirable possibility, but others fear allowing institutional forces to decide which genes should be kept and which discarded.

———. "The Ultimate Therapy." *Tikkun*, vol. 13, May–June 1998, pp. 33ff. Gene therapy is now used only to treat disease, but in future it could be more broadly applied, with unpredictable consequences for civilization and humanity.

Seppa, Nathan. "Gene Therapy Advances Go to the Heart." *Science News*, vol. 154, November 28, 1998, p. 346. Recent research suggests that gene therapy may eventually provide an alternative to some kinds of heart and blood vessel surgery by helping the body grow new blood vessels to replace blocked ones.

Shenk, David. "Biocapitalism: What Price the Genetic Revolution?" *Harper's Magazine*, vol. 294, December 1997, pp. 37ff. Discusses ethical, political, and legal issues raised by recent advances in gene therapy, including abortion, correcting "defects" in fetuses, and attempts to create "designer babies."

Silver, Lee, Jeremy Rifkin, and Barbara Katz Rothman. "Biotechnology: A New Frontier of Corporate Control." *Tikkun*, vol. 13, July–August 1998, pp. 47ff. Roundtable discussion, moderated by Rothman, between a supporter (Silver) and a critic (Rifkin) of genetic alteration of humans; considers the danger of only wealthy parents being able to enhance their children's genetic makeup, thus creating two classes of people.

Taubes, Gary. "Ontogeny Recapitulated." *Discover*, vol. 19, May 1998, pp. 66–72. Describes research on embryonic stem cells and other techniques that could be used to make human tissues and, possibly, organs for transplants.

Travis, John. "Gene Therapy Escapes the Immune Response." *Science News*, vol. 148, September 2, 1995, p. 149. Early attempts at gene therapy were often frustrated by the immune system, which saw the genes as foreign substances and set up a reaction to them; as a result, the therapy could be used only once in a patient. Scientists have now found a way to sneak new genes into the body more than once.

———. "Inner Strength." *Science News*, vol. 153, March 14, 1998, pp. 174–175. Scientists are attempting to use gene therapy to create immune system cells in which HIV cannot replicate, in effect producing an intracellular vaccine against AIDS.

Williams, Gurney III. "Altered States." *American Legion*, vol. 143, October 1997, pp. 23ff. Cloning and other potential changes to human genes could improve human health, but their possible positive and negative consequences must be weighed against the values of society.

Wivel, Nelson A. "Germ-line Gene Modification and Disease Prevention: Some Medical and Ethical Perspectives." *Science*, vol. 262, October 22, 1993, pp. 533–536. Summarizes research on germ-line gene modification in animals and considers ethical issues raised by human germ-line gene modification. Concludes that research in this area should continue so that germ-line modification can be compared with alternate strategies for preventing genetic diseases.

HUMAN CLONING

Note: Articles that discuss Dolly, the cloned sheep, primarily as a precursor of human cloning appear in this section. Articles that discuss Dolly as a scientific advance in connection with animal cloning appear in the "Agricultural Biotechnology" section.

BOOKS

Andrews, Lori B. *The Clone Age: Adventures in the New World of Reproductive Technology*. New York: Holt, 1999. Memoir by an expert on legal and ethical implications of reproductive technology focuses on "a new breed of scientist" propelling technology ahead of legal and ethical ground rules and the wrenching issues that result, including that of possible human cloning.

Annotated Bibliography

Cole-Turner, Ronald, ed. *Human Cloning: Religious Responses*. Louisville, Ky.: Westminster John Knox, 1998. Short essays discuss ethical and religious issues raised by human cloning, both of embryos for research purposes and of potential living children.

Hefley, James C., and Lane P. Lester. *Human Cloning: Playing God or Scientific Progress?* Grand Rapids, Mich.: Fleming H. Revell, 1998. A conservative evangelical viewpoint on the subject, stressing the Bible as a framework within which questions about cloning or any other subject can be answered. Considers artificial insemination, genetic testing, and scientific breeding as well as cloning.

Humber, James M., and Robert Almeder, eds. *Human Cloning (Biomedical Ethics Reviews)*. Humana Press, 1998. Includes different ethical and religious positions on the subject.

Kass, Leon R., and James Q. Wilson. *The Ethics of Human Cloning*. Washington, D.C.: AEI Press, 1998. Authors debate potential moral costs and benefits of human cloning and demonstrate "the wisdom of repugnance" for certain uses of science. Kass is firmly against human cloning, while Wilson thinks it might be acceptable if cloned children are raised in loving, two-parent households.

McCuen, Gary E., ed. *Cloning: Science and Society*. Hudson, Wis.: Gary E. McCuen Publications, 1998. Collection of essays gives pros and cons regarding both animal and human cloning. It includes a section on the legal status of human cloning, a discussion of possible bans on human cloning, and views on ethical and religious aspects of the subject.

McGee, Glenn, ed. *The Human Cloning Debate*. Berkeley, Calif.: Berkeley Hills Books, 1998. Anthology of essays describes the shape of the cloning debate, regulation of cloning, misunderstandings about human cloning, political issues, and religious views.

Marlin, George J. *Politician's Guide to Assisted Suicide, Cloning, and Other Current Controversies*. Morley Institute, 1998. Provides philosophical and historical perspectives on cloning and other difficult ethical issues and argues that without respect for personhood, liberty will become license and responsibility will be abrogated.

Nussbaum, Martha C., and Cass R. Sunstein, eds. *Clones and Clones: Facts and Fantasies About Human Cloning*. New York: W. W. Norton, 1998. Anthology of articles (including four fiction pieces) covers scientific developments in cloning, potential human cloning, and ethical issues raised by the possibility. Contributors include experts in science, sociology, ethics, religion, law, and public policy.

Pence, Gregory E., ed. *Flesh of My Flesh: The Ethics of Cloning Humans, A Reader*. Lanham, Md.: Rowman & Littlefield, 1998. Collection of essays on pros and cons of cloning humans includes short excerpts from the June

1997 report/recommendations of the National Bioethics Advisory Commission.

———. *Who's Afraid of Human Cloning?* Lanham, Md.: Rowman and Littlefield, 1998. Clears up misconceptions and defends the possibility and morality of human cloning.

Rantala, M. L., and Arthur J. Milgram, eds. *Cloning: For and Against.* Chicago: Open Court, 1999. Anthology provides scientific background and offers opposing viewpoints on ethical, religious, and legal issues related to human cloning.

Rorvik, David M. *In His Image: The Cloning of a Man.* Boston: J. B. Lippincott, 1978. This book, which claimed that the author helped an anonymous rich man and a scientist set up a human cloning project in Southeast Asia, caused considerable controversy when it was published. It is generally considered to be a hoax, but it sheds light on abuses that might occur if human cloning becomes possible.

Segal, Nancy L. *Entwined Lives: Twins and What They Tell Us About Human Behavior.* New York: Dutton, 1999. Segal describes results of studies at the Twins Studies Center of California State University at Fullerton, of which she is director. Examining identical twins, who are natural clones, can provide many insights into nature versus nurture, cloning, and other provocative subjects.

Winston, Robert. *The Future of Genetic Manipulation.* London: Phoenix, 1997. Author, a professor of fertility studies at the University of London, explains why he is in favor of research into human cloning, especially the cloning of human tissues as possible transplant sources.

Winters, Paul A., ed. *Cloning: At Issue.* San Diego: Greenhaven Press, 1998. Articles express different viewpoints on cloning, primarily human cloning.

Wright, Lawrence. *Twins: And What They Tell Us About Who We Are.* New York: Wiley, 1997. Identical twins are natural clones, so comparison of the lives of identical twins sheds light, not only on which characteristics are produced by genetics and which by environment, but on differences that might arise between artificially cloned humans.

ARTICLES

Annas, George J. "Why We Should Ban Human Cloning." *New England Journal of Medicine*, vol. 339, July 9, 1998, pp. 122ff. Author feels that human cloning should be banned until it can be regulated to ensure that the technique will be used wisely; he fears that cloned people would be denied their uniqueness.

Bailey, Ronald. "The Twin Paradox: What Exactly Is Wrong with Cloning People?" *Reason*, vol. 29, May 1997, pp. 8–10. Responds to ethical criticisms of possible human cloning, pointing out, among other things, that natural clones already exist in the form of identical twins.

Campbell, Courtney S., and Joan Woolfrey. "Norms and Narratives: Religious Reflections on the Human Cloning Controversy." *Journal of Biolaw and Business*, vol. 1, 1998, pp. 8–20. Presents views on human cloning from various religious communities and concludes that scientists and politicians need to learn more about, and learn from, religious perspectives on this issue.

Cimons, Marlene, and Jonathan Peterson. "Clinton Bans U.S. Funds for Human Cloning Research." *Los Angeles Times*, March 5, 1997, p. A1. Soon after the announcement of the cloning of a sheep, which prompted heated debate about the ethics of possible human cloning, President Clinton banned the use of federal funds for human cloning research and asked scientists in the private sector to refrain from such experiments.

"Cloning Raises Difficult Issues for White House, Congress." *Issues in Science and Technology*, vol. 13, Summer 1997, p. 22. President Clinton and Congress have initiated measures to prevent use of federal funds for research on human cloning and restrict such research in other ways, but several key scientists questioned these measures and testified about possible benefits of cloning technology in a Senate hearing.

Cole, K. C. "Upsetting Our Sense of Self." *Los Angeles Times*, April 28, 1997, pp. A1, A13. Discusses the troubling questions that the prospect of human cloning raises about the meaning and value of individuality; sidebars include quotes about the self and a list of books and movies that feature twins, doubles, or humanlike beings created by scientists.

Easterbrook, Gregg. "Will Homo Sapiens Become Obsolete?" *New Republic*, March 1, 1999, pp. 20ff. Describes recent discoveries about human embryonic stem cells, which potentially can grow into any kind of tissue, and their implications for human embryo research and cloning.

Eibert, Mark D. "Clone Wars." *Reason*, vol. 30, June 1998, pp. 52ff. Author feels that laws banning the scientific or reproductive use of human cloning contradict constitutional principles such as the right to have children and the right of scientists to explore nature.

Fackelmann, Kathy A. "Cloning Human Embryos." *Science News*, vol. 145, February 5, 1994, pp. 92–95. Describes the cloning of human embryos (the embryos were not allowed to develop past an early stage) in 1993, an experiment that created as much controversy in its time as has the more recent cloning of Dolly the sheep.

Garvey, John. "The Mystery Remains: What Cloning Can't Reproduce." *Commonweal*, vol. 124, March 28, 1997, pp. 6–7. Author is opposed to human cloning because it would take away the sacredness and mystery in human procreation, turning people into products.

Harris, Mark. "To Be or Not to Be?" *Vegetarian Times*, no. 250, June 1998, pp. 64ff. Discusses human cloning, which some think may occur before the year 2000, and describes some possible beneficial effects of cloning research, such as production of tissue for burn victims.

Herbert, Wray, Jeffery L. Sheler, and Traci Watson. "The World After Cloning." *U.S. News & World Report*, vol. 122, March 10, 1997, pp. 59ff. The news that an adult sheep was cloned in Scotland raises ethical and religious issues, especially in relation to possible human cloning.

"Human Cloning: Should the United States Legislate Against It?" *ABA Journal*, May 1997, pp. 80–81. George J. Annas, health law professor at Boston University, favors banning human cloning, whereas John A. Robertson, professor of law at the University of Texas, is opposed to a ban.

Kass, Leon R. "The Wisdom of Repugnance: Why We Should Ban the Cloning of Humans." *New Republic*, vol. 216, June 2, 1997, pp. 17ff. Long, thoughtful essay maintains that people should trust their repugnance toward human cloning, which challenges values such as those respecting sex, the body, and individuality and could turn reproduction into mere manufacturing.

Kassirer, Jerome P., and Nadia A. Rosenthal. "Should Human Cloning Research Be Off Limits?" *New England Journal of Medicine*, vol. 338, March 26, 1998, pp. 905–906. In this editorial, the *NEJM* adds its voice to those of scientists and biotechnology companies who oppose legislation that would ban all research on cloning of human cells. The editors feel that cloned cells have uses in medicine, but they are willing to accept a bill that bans implantation of a cloned embryo in a human uterus.

Kilner, John F. "Stop Cloning Around." *Christianity Today*, vol. 41, April 28, 1997, pp. 10–11. Author claims that human cloning is undesirable because each person is a unique creation of God, not a product, and that research on human cloning is immoral because it would cause the death of human beings (embryos).

Kolata, Gina. "Cloning Human Embryos: Debate Erupts Over Ethics." *The New York Times*, October 26, 1993, pp. A1, B7. Describes ethical questions raised by the cloning of early-stage human embryos by scientists at the George Washington University Medical Center.

———. "For Some Infertility Experts, Human Cloning Is a Dream." *The New York Times*, June 7, 1997, p. A6. Although critics have denounced possible human cloning, infertility experts such as Mark Sauer of Columbia Presbyterian Medical Center in New York point out that it may be the answer to a prayer for some infertile couples.

Kluger, Jeffrey. "Will We Follow the Sheep?" *Time*, vol. 149, March 10, 1997, pp. 66–71. The recent cloning of a sheep raises the prospect that humans might also be cloned in the near future, which presents major eth-ical problems. Article describes President Clinton's call for study of the issue and presents results of an opinion poll on cloning.

Kontorovich, E. V. "Asexual Revolution." *National Review*, vol. 50, March 9, 1998, pp. 30ff. Sees human cloning as dangerous because it would blur parental heritage and remove clones' essential humanity; claims that pro-posals such as harvesting organs from aborted fetuses illustrate modern society's moral decline.

Lefevre, Patricia. "Cloning Raises Ethical Questions About Life, Human Limits and Love." *National Catholic Reporter*, vol. 33, March 14, 1997, p. 5. Discusses religious objections to human cloning, which are based on the right of humans to be procreated in a human way rather than in a lab-oratory, and presents comments from church representatives.

LeVay, Simon. "Darwin's Children." *Advocate*, no. 744, October 14, 1997, pp. 63–64. Human cloning could benefit homosexuals by allowing homosex-ual couples to have children.

Lewontin, R. C. "The Confusion over Cloning." *New York Review*, October 23, 1997, pp. 18–22. Summarizes and comments on *Cloning Human Beings*, the 110-page report of the National Bioethics Advisory Commission.

Marshall, Eliot. "Biomedical Groups Derail Fast-Track Anticloning Bill." *Science*, vol. 279, February 20, 1998, pp. 1123. Showing unusual lobbying savvy and political muscle, biomedical research groups halted legislation aimed at banning human cloning that was moving on a fast track through the U.S. Senate.

Mautner, Michael. "Will Cloning End Human Evolution?" *The Futurist*, vol. 31, November–December 1997, p. 68. Because cloning foregoes the genetic mixing inherent in sexual reproduction, it could stop human evolution if it were to become the primary mode of human reproduc-tion.

Motavalli, Jim, and Tracey C. Rembert. "Me and My Shadow." *E Magazine*, vol. 8, July–August 1997, pp. 15–21. Describes environmental and ethical issues raised by the recent cloning of a sheep and the possibility of human cloning and quotes a variety of opinions about those issues.

Nash, J. Madeleine. "Cloning's Kevorkian." *Time*, vol. 151, January 19, 1998, p. 58. Brief interview with Richard Seed, eccentric physicist and advocate of human cloning.

Post, Stephen G. "The Judaeo-Christian Case Against Cloning." *America*, vol. 176, June 21, 1997, pp. 19ff. Analyzes seven bioethical arguments against human cloning in terms of Biblical Christian traditions. Claims that cloning technology typifies confusion of means and ends in modern society.

Riddell, Mary. "Just Because the Idea of Cloning Provokes Outrage Doesn't Mean It Should Be Banned." *New Statesman*, vol. 127, December 11, 1998, p. 10. Britain's Human Genetics Advisory Commission has recommended a "total ban" on human cloning but advises permitting use of very young embryos for research on tissue cloning, which might produce treatments for illnesses such as Parkinson's disease.

Robertson, John A. "Human Cloning and the Challenge of Regulation." *New England Journal of Medicine*, vol. 339, July 9, 1998, pp. 119ff. Author believes that human cloning should be regulated to ensure responsible use but not banned; it could be a valid reproductive choice for some couples, and experience with identical twins shows that having identical DNA does not deny individuality or freedom of expression.

———. "Liberty, Identity, and Human Cloning." *Texas Law Review*, vol. 76, June 1998, pp. 1371–1456. Discusses human cloning in relation to reproductive liberty and concludes that potential harms of cloning should be weighed against respect for procreative liberty and autonomy.

Stiefel, Chana Freiman. "Cloning: Good Science or Baaad Idea?" *Science World*, vol. 53, May 2, 1997, pp. 8ff. Describes history of animal and human cloning efforts and bans on human cloning research.

"To Clone or Not to Clone?" *Christian Century*, vol. 114, March 19, 1997, pp. 286ff. Some religious and political groups expressed vigorous objections to possible human cloning in the wake of news about the cloning of a sheep, but other religions had a wait-and-see attitude.

"Uproar over Cloning." *Christian Century*, vol. 115, January 28, 1998, pp. 76–77. The Council of Europe agreed on January 12 to ban human cloning as a threat to human dignity.

Van Gelder, Lindsay. "Hello, Dolly, Hello, Dolly." *Ms. Magazine*, vol. 7, May–June 1997, p. 26. Maintains that human cloning would empower women by freeing them of a need to be with men in order to reproduce, but notes that society is not yet supportive of this idea and thus has halted research on human cloning.

Verhey, Allen. "Theology After Dolly: Cloning and the Human Family." *Christian Century*, vol. 114, March 19, 1997, pp. 285–286. The ethics of human cloning has been debated for more than 30 years, but the debate took on new fervor following news that a sheep had been cloned. Supporters say people should have the freedom to make their own choices in the matter and that cloning good people could help society.

Wachbroit, Robert. "Genetic Encores: The Ethics of Human Cloning." *Report from the Institute for Philosophy and Public Policy*, vol. 17, Fall 1997, pp. 1–7. Discusses interests and rights, concerns about process, and reasons for cloning; considers that other techniques of human genetic engineering are likely to advance faster than cloning and will prevent human cloning from becoming widespread.

Wade, Nicholas. "Ethics Panel OKs Research on Cells from Human Embryos." *The New York Times*, reprinted in *San Francisco Chronicle*, June 29, 1999. The National Bioethics Advisory Commission has recommended that federally financed scientists be allowed to continue research on stem cells taken from human embryos because the research offers such great promise. Human embryos might be cloned to provide stem cells for research.

Wilmut, Ian. "Dolly's False Legacy." *Time*, vol. 153, January 11, 1999, p. 74. The Scottish scientist who created Dolly, the first mammal cloned from a mature adult cell, lists scientific and ethical questions raised by possible human cloning.

Zindler, Frank R. "Spirits, Souls, and Clones: Biology's Latest Challenge to Theology." *American Atheist*, Summer 1997, pp. 17–26. Discusses relationship of views of the soul to religious opposition to human cloning and claims that "cloning of humans . . . would be the reductio ad absurdum of the theological notion of souls."

INTERNET/WEB DOCUMENTS

Andrews, Lori B. "Mom, Dad, Clone: Why We Shouldn't Build Families Through Human Cloning." Available online. URL: http://www.law.uh.edu/Health/HealthLawNews/06-1998.html. Posted June 1998. Article from *Health Law News* maintains that the right to bear cloned children should not be protected by the Constitution because of the physical and psychological risks to the child.

Human Cloning Foundation. "The Benefits of Human Cloning." Available online. URL: http://www.humancloning.org/benefits.htm. Posted 1998. Offers reasons why human cloning could be beneficial and circumstances under which it might be ethically used.

National Bioethics Advisory Commission. *Cloning Human Beings*. Available online. URL: http://www.bioethics.gov/pubs.html. Posted June 1997. Report on the science and ethics of human cloning, commissioned by President Clinton in the wake of the news that a sheep had been cloned from an adult body cell; includes options for regulation. Recommends a law imposing a three-to-five-year moratorium on producing children by cloning.

CHAPTER 8

ORGANIZATIONS
AND AGENCIES

There are many organizations devoted to biotechnology/genetic engineering and various aspects of genetics. The following entries include professional groups, industry organizations, advocacy groups, and government agencies, both in the United States and abroad. In keeping with the widespread use of the Internet and e-mail, the web site (URL) address and e-mail address are given first when available, followed by the phone number and postal address.

Alliance for Bio-Integrity
URL: www.bio-integrity.org/index.html
E-mail: info@bio-integrity.org
Phone: (515) 472-5554
406 West Depot Avenue
Fairfield, IA 52556
Believes that genetically engineered foods present "unprecedented dangers to the environment and human health" and should be labeled and carefully regulated.

Alliance of Genetic Support Groups
URL: http://www.healthy.net/pan/cso/cioi/AGSG.htm
Phone: (800) 336-4363

4301 Connecticut Avenue NW, Suite 404
Washington, DC 20008
Consortium of organizations for families and individuals suffering from genetic disorders. Its concerns include possible discrimination against people with genetic defects or their families.

American Civil Liberties Union
URL: http://www.aclu.org
E-mail: aclu@aclu.org
Phone: (212) 549-2500
125 Broad Street, 18th Floor
New York, NY 10004-2400
One area of concern for this non-profit, nonpartisan public interest

organization devoted to protecting American civil liberties is genetic privacy. It opposes employment and insurance discrimination based on genetic testing as well as the establishment of national DNA databases. It offers both print and online resources, including books, pamphlets, newsletters, and web links.

American College of Medical Genetics
URL: http://www.aai.org/genetics/acmg/acmgmenu.htm
E-mail: acmg@faseb.org
Phone: (301) 530-7127
9650 Rockville Pike
Bethesda, MD 20814-3998
Professional organization for scientists and health care professionals specializing in medical genetics. Among other things, provides guidelines for genetic health testing and lobbies for effective and fair health policies and legislation.

American Crop Protection Association
URL: http://www.acpa.org/index.html
E-mail: member-services@acpa.org
Phone: (202) 296-1585
1156 Fifteenth Street, N.W., Suite 400
Washington, DC 20005
Represents the agricultural pesticide industry and supports its products, including bioengineered plants containing Bt or other pesticide genes.

American Genetics Association
Phone: (301) 695-9292
P. O. Box 257
Buckeystown, MD 21717-0257
Promotes basic and applied research on the genetics of plants and animals.

American Society for Reproductive Medicine
URL: http://www.asrm.org/
E-mail: asrm@asrm.org
Phone: (205) 978-5000
1209 Montgomery Highway
Birmingham, AL 35216-2809
Professional organization devoted to advancing knowledge and expertise in reproductive medicine and biology. Publishes educational and other materials, including some that deal with ethical considerations.

American Society of Gene Therapy
URL: http://www.asgt.org
E-mail: asgt@slackinc.com
Phone: (856) 848-1000
6900 Grove Road
Thorofare, NJ 08086-9447
Fosters education, exchange of information, and research on gene therapy.

American Society of Human Genetics
URL: http://www.faseb.org/genetics/ashg/ashgmenu.htm
Phone: (301) 571-1825
9650 Rockville Pike
Bethesda, MD 20814-3998
Professional society of researchers, physicians, genetic counselors and others interested in human genetics

and related social issues. Publishes monthly *American Journal of Human Genetics*.

American Society of Law, Medicine, and Ethics
URL: http://www.aslme.org
Phone: (617) 262-4990
765 Commonwealth Avenue, 16th Floor
Boston, MA 02215
Members include attorneys, physicians, health care administrators, and others interested in relationship between law, medicine, and ethics. Publishes *American Journal of Law* and *Journal of Law, Medicine, and Ethics*, both quarterly.

Bioengineering Action Network of North America (BAN)
URL: http://www.tao.ca/~ban
E-mail: ban@tao.ca
Activist group strongly opposed to bioengineered crops and food.

Biotechnology Industry Organization
URL: http://www.bio.org
Phone: (202) 857-0244
1625 K Street, N.W., #1100
Washington, DC 20006
Chief trade and lobbying organization for the biotechnology industry, including academic institutions as well as commercial companies.

Biotechnology Working Group
URL: http://www.edf.org/pubs/ Reports/bwg.html
E-mail: EDF@edf.org
Phone: (800) 684-3322

257 Park Avenue South
New York, NY 10010
Associated with the Environmental Defense Fund and a project of the Tides Foundation, the Biotechnology Working Group tries to strengthen the influence of the public interest community on the development of biotechnology.

British Society for Human Genetics
URL: http://www.bham.ac.uk/ BSHG/
E-mail: bshg@bham.ac.uk
Phone: (44) 121 627-2634
Clinical Genetics Unit
Birmingham Women's Hospital
Birmingham B15 2TG
England
Professional association for British scientists and health professionals involved in human genetics. Its Public Policy Committee considers the ethical and social effects of human genetic research and technology. It publishes a quarterly newsletter and issues statements on aspects of human genetics.

Centre for Applied Bioethics
URL: http://www.nottingham.ac. uk/bioethics/
E-mail: ben.mepham@ nottingham.ac.uk
Phone: (044 115) 951-6303
School of Biological Sciences
University of Nottingham
Sutton Bonington Campus
Loughborough
Leicester LE12 5RD
England

Concerned with appropriate application of biotechnology to food production, industry, and medical uses of farm animals.

Clone Rights United Front
Clone Rights Action Center
URL: http://www.humancloning.
 org/users/randy/index.html
E-mail: randy@www.
 humancloning.org
Phone: (212) 255-1439
506 Hudson Street
New York, NY 10014
Claims that the right to be cloned is part of the right to control reproduction. Opposes bans on human cloning.

Council for Responsible Genetics
URL: http://www.gene-watch.org
E-mail: crg@essential.org
Phone: (617) 868-0870
5 Upland Road, Suite 3
Cambridge, MA 02140
National nonprofit organization of scientists, public health professionals, and others that works to see that biotechnology develops safely and in the public interest. Its areas of concern include genetic discrimination, patenting of life-forms, food safety, and environmental quality. Publishes a bimonthly newsletter, GeneWATCH, and educational materials.

Council of Regional Networks
 (CORN)
URL: http://www.kumc.edu/gec
Phone: (404) 727-1475
Emory University School of
 Medicine

Department of
 Pediatrics/Genetics
2040 Ridgewood Drive
Atlanta, GA 30322
Association of regional genetics networks; provides list of members.

Cultural Survival
URL: http://www.cs.org/
E-mail: csinc@cs.org
Phone: (617) 441-5400
96 Mount Auburn Street
Cambridge, MA 02138-5017
Defends human rights and cultural autonomy of indigenous peoples and oppressed ethnic minorities, including protection from exploitation by multinational biotechnology companies. Publishes a journal, *Cultural Survival Quarterly*.

Department of Agriculture
 (USDA)
Animal and Plant Health
 Inspection Service (APHIS)
URL: http://www.aphis.usda.gov/
E-mail: john.t.turner@usda.gov
Phone: (301) 734-7601
Biotechnology Evaluation
USDA-APHIS-PPQ
4700 River Road, Unit 147
Riverdale, MD 20737-1236
United States government agency that regulates any organisms, including genetically altered organisms, that are or might be plant pests. Offers publications on genetically engineered plants and related subjects.

Environmental Protection Agency (EPA)
URL: http://www.epa.gov/
E-mail: hq.faia@epa.mail.epa.gov
Phone: (201) 260-4048
401 M Street SW (1105)
Washington, DC 20460
United States government agency that regulates pesticides, including biopesticides, and sets limits for the amounts of such substances that can remain on or in food; also regulates new chemicals, which are considered to include some genetically engineered organisms.

Eugenics Special Interest Group
Phone: (518) 732-2390
P. O. Box 138
East Schodack, NY 12063
Promotes a positive view of eugenics.

European Federation of Biotechnology
URL: http://sci.mond.org/efb
E-mail: efb.cbc@stm.tudelft.nl
Phone: (44) 171 235-3681
14/15 Belgrave Square
London SW1X 8PS
England
Works to increase public and government understanding of biotechnology. Publications include a newsletter and briefing papers on patenting life, biotechnology in foods and drinks, the application of human genetic research, and environmental technology.

European Molecular Biology Organization (EMBO)
URL: http://www.embo.org

E-mail: EMBO@EMBL-Heidelberg.de
Phone: (49) 6221 383031
Postfach 1022.40
D-69012 Heidelberg
Germany
Established in 1962, EMBO promotes molecular biology in Europe and neighboring countries. Publishes the *EMBO Journal*. Its Science and Society Committee communicates with the nonscientific community about the effects and benefits of molecular biology. It sponsors research and training through the European Molecular Biology Laboratory and its outstations.

European Society of Human Genetics
URL: http://www.eshg.org
E-mail: eshg@ ehsg.org
Phone: (44) 121 623-6830
Clinical Genetics Unit
Bitrmingham Women's Hospital
Birmingham B15 2TG
England
Arranges meetings and other scientific gatherings in the field of human genetics and publishes *The European Journal of Human Genetics*.

Food and Drug Administration (FDA)
URL: http://www.fda.gov
E-mail: webmail@oc.fda.gov
Phone: (888) 463-6332
Office of Consumer Affairs
5600 Fishers Lane, Room 1685
Rockville, MD 20857
United States government agency that regulates food, including genet-

ically modified foods. Web site has a page on bioengineered food.

Food First/Institute for Food and Development Policy
URL: www.foodfirst.org
E-mail: foodfirst@foodfirst.org
Phone: (800) 274-7826
398 60th Street
Oakland, CA 94618
Think tank highlights root causes and value-based solutions to world hunger and poverty. Opposes control of agricultural biotechnology and food supply by large multinational corporations.

Foundation on Economic Trends
URL: www.biotechcentury.org
E-mail: jrifkin@foet.org
Phone: (202) 466-2823
1660 L Street, N.W., Suite 216
Washington, DC 20036
Headed by Jeremy Rifkin, this nonprofit foundation examines emerging trends in science and technology and their impacts on society, culture, the economy, and the environment. Areas of interest include transgenic animals, patents on living things, and human gene alteration. The group urges caution in use of genetic technology.

Friends of the Earth
URL: http://www.foe.org/
E-mail: foe@foe.org
Phone: (202) 783-7400
1025 Vermont Avenue, N.W.
Washington, DC 20005-6303
Part of an international federation of environmental organizations (Friends of the Earth International) that try to work toward sustainable societies, protect the environment (including biological diversity), and promote justice and equal access to resources and opportunities for all the world's people. Among other things, the group fears the effects of multinational corporations' control of genetically modified crops on farmers in developing nations.

Genetic Interest Group
URL: http://www.gig.org.uk/
Phone: (44) 0171 704 3141
Unit 4d, Leroy House
436 Essex Road
London N1 3QP
Britain
Alliance of British support groups for people and families with inherited diseases or diseases with a significant genetic component. Among other things, the group works to prevent discrimination based on the misapplication of genetic information, to promote recognition of the health benefits that can result from genetic research, and to guarantee that such research ultimately benefits individual patients.

Genetical Society
URL: http://www.genetics.org.uk
E-mail: gensoc.memsec@bbsrc. ac.uk
Roslin Institute
Roslin, Midlothian EH25 9PS
Britain
Professional society of British geneticists; organizes meetings and

217

journals in the field and publishes a newsletter for members three times a year.

Genetics Society of America
URL: http://www.faseb.org/
  genetics/gsa/gsa-int.htm
E-mail: estrass@genetics.faseb.
  org
Phone: (301) 571-1825
9650 Rockville Pike
Bethesda, MD 20814-3998
Professional society of scientists and academicians working in the field of genetic studies. Publishes monthly journal, *Genetics*, and educational/career materials.

Genome Action Coalition
URL: http://www.tgac.org
E-mail: Idennis@dc-ord.com
Alliance of groups that believe that the success of the Human Genome Project is critically important for biomedical research and health care. Addresses Congress on issues of concern, including protection of medical privacy and prevention of genetic discrimination in insurance and employment.

Greenpeace
URL: http://www.greenpeaceusa.
  org
Phone: (800) 326-0959
1436 U Street NW
Washington, DC 20009
Environmental group whose aims include protection of global biodiversity, requirement of careful regulation and labelling of genetically engineered foods, and opposition to patenting of living things.

Hastings Center
URL: http://www.hastingscenter.
  org/center.htm
Phone: (914) 424-4040
Garrison, NY 10524-5555
The center addresses fundamental ethical issues in health, medicine, and the environment, including issues related to biotechnology and human genetics. Publishes bimonthly journal, *The Hastings Center Report*, a study of ethical issues related to research on human subjects, and other papers.

Health Insurance Association of America
URL: http://www.hiaa.org
Phone: (202) 824-1600
555 13 Street, N.W.
Washington, DC 20004
Advocacy group for the health insurance industry. Opposes legislation that restricts insurers' use of genetic information.

Human Cloning Foundation
URL: http://www.humancloning.
  org
E-mail: HCloning@aol.com
Suite 410A-143
Atlanta, GA 30328
Stresses the positive aspects of human cloning and promotes education, awareness, and research about human cloning and other biotechnology. Web site offers papers supporting human cloning, lists of

books about cloning and genetic engineering, and aids for students.

Human Genetics Advisory Commission
URL: http://www.dti.gov.uk/hgac
E-mail: mileva.novkovic@ osct.dti.gov.uk
Phone: (44) 0171 271 2131
Office of Science and Technology
Albany House
94-98 Petty France
London SW1H 9ST
Britain
Established in 1996 to offer the British government independent advice on issues arising from developments in human genetics. Has published papers (available online) about human genetics research, human cloning, and genetic testing in relation to insurance.

Human Genome Diversity Project
URL: http://www.stanford.edu/ group/morrinst/HGDP.html
Stanford University
Morrison Institute for Population and Resource Studies
Stanford, CA 94305
Project aims to gather DNA samples from a number of small indigenous populations and study them to learn more about such things as differences in resistance to disease. Offers literature on the project.

Human Genome Organization (HUGO)
URL: http://www.gene.ucl.ac.uk/ hugo
E-mail: hugo@hugo-international. org
Phone: (44) 171 935 8085
142-144 Harley Street
London W1N 1AH
England
A leading international (chiefly European) professional organization for scientists who study the human genome.

The Innocence Project (affiliated with the National Association of Criminal Defense Lawyers)
URL: http://www.criminaljustice. org:80/PUBLIC/innocent.htm
E-mail: assist@nacdl.com
Phone: (202) 872-8600
1025 Connecticut Avenue N.W., Suite 901
Washington, DC 20036
Founded by NACDL members Barry Scheck and Peter Neufeld at the Cardozo Law School in New York, the project uses DNA tests and other evidence to show that certain people convicted of crimes are innocent and obtain their release.

Institute for Agriculture and Trade Policy
URL: http://www.iatp.org/iatp
E-mail: khoff@iatp.org
Phone: (612) 870-0453
2105 First Avenue South
Minneapolis, MN 55404

Nonprofit research and education organization aimed at creating environmentally and economically sustainable communities and regions through sound agriculture and trade policy. Prepares educational materials for policymakers and the public. Publishes free weekly news-letter.

The Institute for Genomic Research (TIGR)
URL: http://www.tigr.org
Phone: (301) 838-0200
9712 Medical Center Drive
Rockville, MD 20850
TIGR is a nonprofit research institute that analyzes the genomes and gene products of a wide variety of organisms, from viruses to humans.

International Center for Technology Assessment
URL: http://www.icta.org/aboutus/index.htm
E-mail: office@icta.org
Phone: (202) 547-9359
310 D Street NE
Washington, DC 20002
Analyzes impacts of technology on society. Concerns include limiting genetic engineering, halting the patenting of life, and defending the integrity of food. Sponsors the Campaign for Food Safety and the Biotechnology Watch Human Applications.

International Centre for Genetic Engineering and Biotechnology
URL: http://base.icgeb.trieste.it/
E-mail: kerbav@icgeb.trieste.it
Phone: (39) 040 37571
AREA Science Park

Padriciano 99
34012 Trieste
Italy
Established by the United Nations Industrial Development Organization in 1995 to provide research, training, and scientific services in the safe use of molecular biology and biotechnology, especially as it applies to the needs of developing nations.

International Cloning Society
URL: http://www.angelfire.com/la/;fled/ics.html
E-mail: adlafferty@msn.com
Acts as agency to preserve cell/DNA specimens of people who want to be cloned in the future, particularly for purposes of space travel.

International Federation of Human Genetics Societies
URL: http://www.faseb.org/genetics/international
E-mail: estrass@faseb.org
Phone: (301) 571-1825
9650 Rockville Pike
Bethesda, MD 20814-3998
Provides a forum for groups dedicated to all aspects of human genetics and transmits policy statements from the federation to appropriate parties.

International Food Information Council
URL: http://ificinfo.org/
E-mail: foodinfo@ificinfo.health.org
1100 Conecticut Avenue, N.W., Suite 430
Washington, DC 20036

Communicates science-based information on food safety and nutrition to health professionals, media, and others providing information to consumers. Supported by the food, beverage, and agricultural industries.

**International Food Policy
    Research Institute**
URL: http://www.cigar.org/ifpri/
    index.htm
E-mail: fpri@cigar.org
Phone: (202) 862-5600
2033 K Street, N.W.
Washington, DC 20006-1002
Identifies and analyzes national and international strategies and policies for meeting food needs of the developing world in a sustainable way.

**International Genetics
    Federation (IGF)**
URL: http://www.igf.org
Phone: (604) 822-5629
University of British Columbia,
    Department of Botany
6270 University Boulevard
Vancouver, British Columbia
    V6T 1Z4
Canada
The IGF is a federation of the world's genetical societies. Its purpose is to promote the science of genetics at the international level.

**International Society for
    Environmental Biotechnology**
URL: http://cape.uwaterloo.ca/
    research/iseb/iseb.htm
E-mail: Wanderson@uwaterloo.ca

Phone: (519) 888-4567
University of Waterloo
Waterloo, Ontario N2L 3G1
Canada
Interdisciplinary federation of scientists and others interested in environmental biotechnology, or the development, use, and regulation of biological systems for remediation of contaminated environments and for environment-friendly processes. Headquartered in Canada.

**Kennedy Institute of Ethics
    (Georgetown University)**
URL: http://www.georgetown.
    edu /research/kie
E-mail: kicourse@gunet.
    georgetown.edu
Phone: (202) 687-8099
Box 571212
Washington, DC 20057-1212
This teaching and research center sponsors research on medical ethics and related policy issues, including issues related to human genetics and gene alteration. Publishes a journal, newsletter, a bibliography, and other materials.

**March of Dimes Birth Defects
    Federation**
URL: http://www.modimes.org
Phone: (888) 663-4637
1275 Mamaroneck Avenue
White Plains, NY 10605
Works to prevent and treat birth defects, including those caused by genetic mutations. Offers educational material on genetic and other birth defects and a quarterly newsletter, *Genetics in Practice*.

Ministry of Agriculture,
 Fisheries, and Food
URL: http://www.dti.gov.uk/CB/
 bioguide/maff.htm#contents
Phone: (44) 171 238-6377
Additives and Novel Foods
 Division
Ergon House, c/o Nobel House
17 Smith Square
London SW1P 3JR
England
British government agency that works to support the development of food and agricultural biotechnology while protecting people, livestock, crops, and the natural environment. It is partly responsible for regulating pesticides, food products, and plant and animal health and consults on release of genetically modified organisms. Offers publications on the use of biotechnology, including genetically modified foods.

Mothers for Natural Law
URL: http://www.safe-food.org
E-mail:mothers@natural-law.
 org
Phone: (515) 472-2040
P.O. Box 1900
Fairfield, IA 52556
Opposes genetic engineering, especially genetically engineered food.

National Action Plan on Breast
 Cancer
URL: http://www.napbc.org/
 Default.htm
E-mail: napbcinfo@hq.row.com
Phone: (202) 401-9587
U.S. Public Health Service
Office on Women's Health

H. H. Humphrey Building,
 Room 718F
200 Independence Avenue
Washington, DC 20201
Established in 1994 to develop a comprehensive strategy to reduce the incidence of breast cancer. Its Hereditary Susceptibility Working Group focuses on women who have inherited genes that make them unusually susceptible to breast cancer and offers several publications on this subject, including ones on genetic discrimination in employment.

National Agricultural
 Biotechnology Council
URL: http://www.cals.cornell.
 edu/extension/nabc
E-mail: NABC@cornell.edu
Phone: (607) 254-4856
419 Boyce Thompson Institute
Tower Road
Ithaca, NY 14853
Provides an open forum to discuss issues related to agricultural biotechnology and encourage the field's safe, ethical, efficacious, and equitable development. Composed of major nonprofit agricultural biotechnology research and/or teaching institutions in Canada and the United States. Offers reports and a newsletter, *NABCnews*.

National Bioethics Advisory
 Commission
URL: http://www.bioethics.gov
Phone: (301) 402-4242
6100 Executive Boulevard
 Suite 5B01
Rockville, MD 20892-7508

Established in 1995 to advise the President on bioethics issues, the National Bioethics Advisory Commission became famous for its June 1997 report on human cloning. It has also written about the use of cells and tissues and the rights of human subjects in biomedical research.

**National Center for
  Biotechnology Information**
URL: http://130.14.22.107/
E-mail: info@ncbi.nim.nih.gov
Phone: (301) 496-2475
**National Library of Medicine
Building 38A, Room 8N805
Bethesda, MD 20894**
Offers information about biotechnology, including a newsletter, *NCBI News*.

**National Council on Gene
  Resources**
Phone: (510) 524-8973
**1738 Thousand Oaks Boulevard
Berkeley, CA 94707**
Promotes world gene resource conservation and protection of biodiversity.

**National Human Genome
  Research Institute (NHGRI)**
URL: www.nhgri.nih.gov/
URL for ELSI Working Group:
  www.nhgri.nih.gov/
Part of the National Institutes of Health, the NHGRI heads the Human Genome Project. It offers reports and databases of genetic sequencing and other research, including links to other groups

doing genome research. Its Ethical, Legal, and Social Implications (ELSI) Working Group has studied such issues as privacy and fairness in use of genetic information. It offers reports and fact sheets on its work.

**National Organization for Rare
  Disorders (NORD)**
URL: www.pcnet.com/~orphan
Phone: (800) 999-6673
**P. O. Box 8923
New Fairfield, CT 06812-8923**
A federation of voluntary health organizations dedicated to helping people with rare ("orphan") diseases and to assisting the organizations that serve them. Publishes a newsletter, *Orphan Disease Update*, and offers several databases online.

**National Society of Genetic
  Counselors**
URL: http://www.nsgc.org/
E-mail: nsgc@aol.com
Phone: (610) 872-7608
**233 Canterbury Drive
Wallingford, PA 19086-6617**
Professional organization of genetic counselors.

**Rural Advancement Foundation
  International (RAFI)**
URL: http://www.rafi.ca/
E-mail: rafi@rafi.org
Phone: (204) 453-5259
**110 Osborne Street, Suite 202
Winnepeg, MB R3L 1Y5**
Supports rights of indigenous people and preservation of biodiversity and natural resources; opposes patenting of living things and exploitation of

indigenous peoples and resources by multinational biotechnology corporations. Has led campaigns against "biopiracy," "Terminator" genes, and patenting of cells and genes of native peoples. Offers publications on these subjects.

Union of Concerned Scientists
URL: http://www.ucusa.org/

E-mail: ucs@ucsusa.org
Phone: (617) 547-5552
Two Brattle Square
Cambridge, MA 02238-9105
Works to improve the environment and protect human health, safety, and quality of life. Opposes most forms of agricultural biotechnology and publishes a newsletter, *The Gene Exchange*, that monitors this industry.

PART III

APPENDICES

APPENDIX A

DIAMOND V. CHAKRABARTY 447 U.S. 303 (1980)

DIAMOND, COMMISSIONER OF PATENTS AND TRADEMARKS V. CHAKRABARTY CERTIORARI TO THE UNITED STATES COURT OF CUSTOMS AND PATENT APPEALS. NO. 79-136

Argued March 17, 1980
Decided June 16, 1980

Title 35 U.S.C. 101 provides for the issuance of a patent to a person who invents or discovers "any" new and useful "manufacture" or "composition of matter." Respondent filed a patent application relating to his invention of a human-made, genetically engineered bacterium capable of breaking down crude oil, a property which is possessed by no naturally occurring bacteria. A patent examiner's rejection of the patent application's claims for the new bacteria was affirmed by the Patent Office Board of Appeals on the ground that living things are not patentable subject matter under 101. The Court of Customs and Patent Appeals reversed, concluding that the fact that microorganisms are alive is without legal significance for purposes of the patent law.

Held:

1. A live, human-made micro-organism is patentable subject matter under 101. Respondent's micro-organism constitutes a "manufacture" or "composition of matter" within that statute. Pp. 308–318.

(a) In choosing such expansive terms as "manufacture" and "composition of matter," modified by the comprehensive "any," Congress contemplated that the patent laws should be given wide scope, and the relevant legislative history also supports a broad construction. While laws of nature, physical phenomena, and abstract ideas are not patentable, respondent's claim is not to a hitherto unknown natural phenomenon, but to a nonnaturally occurring manufacture or composition of matter—a product of human ingenuity "having a distinctive name, character [and] use." *Hartranft v. Wiegmann*, 121 U.S. 609, 615. *Funk Brothers Seed Co. v. Kalo Inoculant Co.*, 333 U.S. 127, distinguished. Pp. 308–310.

(b) The passage of the 1930 Plant Patent Act, which afforded patent protection to certain asexually reproduced plants, and the 1970 Plant Variety Protection Act, which authorized protection for certain sexually reproduced plants but excluded bacteria from its protection, does not evidence congressional understanding that the terms "manufacture" or "composition of matter" in 101 do not include living things. Pp. 310–314. [447 U.S. 303, 304].

(c) Nor does the fact that genetic technology was unforeseen when Congress enacted 101 require the conclusion that micro-organisms cannot qualify as patentable subject matter until Congress expressly authorizes such protection. The unambiguous language of 101 fairly embraces respondent's invention. Arguments against patentability under 101, based on potential hazards that may be generated by genetic research, should be addressed to the Congress and the Executive, not to the Judiciary. Pp. 314–318.

596 F.2d 952, affirmed.

BURGER, C. J., delivered the opinion of the Court, in which STEWART, BLACKMUN, REHNQUIST, and STEVENS, JJ., joined. BRENNAN, J., filed a dissenting opinion, in which WHITE, MARSHALL, and POWELL, JJ., joined, post, p. 318.

Deputy Solicitor General Wallace argued the cause for petitioner. With him on the brief were Solicitor General McCree, Assistant Attorney General Shenefield, Harriet S. Shapiro, Robert B. Nicholson, Frederic Freilicher, and Joseph F. Nakamura.

Edward F. McKie, Jr., argued the cause for respondent. With him on the brief were Leo I. MaLossi, William E. Schuyler, Jr., and Dale H. Hoscheit.*

MR. CHIEF JUSTICE BURGER delivered the opinion of the Court.

We granted certiorari to determine whether a live, human-made micro-organism is patentable subject matter under 35 U.S.C. 101.

I

In 1972, respondent Chakrabarty, a microbiologist, filed a patent application, assigned to the General Electric Co. The application asserted 36 claims related to Chakrabarty's invention of "a bacterium from the genus *Pseudomonas* containing therein at least two stable energy-generating plasmids, each of said plasmids providing a separate hydrocarbon degradative

228

pathway." This human-made, genetically engineered bacterium is capable of breaking down multiple components of crude oil. Because of this property, which is possessed by no naturally occurring bacteria, Chakrabarty's invention is believed to have significant value for the treatment of oil spills.

Chakrabarty's patent claims were of three types: first, process claims for the method of producing the bacteria; [447 U.S. 303, 306] second, claims for an inoculum comprised of a carrier material floating on water, such as straw, and the new bacteria; and third, claims to the bacteria themselves. The patent examiner allowed the claims falling into the first two categories, but rejected claims for the bacteria. His decision rested on two grounds: (1) that micro-organisms are "products of nature," and (2) that as living things they are not patentable subject matter under 35 U.S.C. 101.

Chakrabarty appealed the rejection of these claims to the Patent Office Board of Appeals, and the Board affirmed the examiner on the second ground. Relying on the legislative history of the 1930 Plant Patent Act, in which Congress extended patent protection to certain asexually reproduced plants, the Board concluded that 101 was not intended to cover living things such as these laboratory created micro-organisms.

The Court of Customs and Patent Appeals, by a divided vote, reversed on the authority of its prior decision in *In re Bergy*, 563 F.2d 1031, 1038 (1977), which held that "the fact that microorganisms . . . are alive . . . [is] without legal significance" for purposes of the patent law. Subsequently, we granted the Acting Commissioner of Patents and Trademarks' petition for certiorari in *Bergy*, vacated the judgment, and remanded the case "for further consideration in light of *Parker v. Flook*, 437 U.S. 584 (1978)" 438 U.S. 902 (1978). The Court of Customs and Patent Appeals then vacated its judgment in *Chakrabarty* and consolidated the case with *Bergy* for reconsideration. After re-examining both cases in the light of our holding in *Flook*, that court, with one dissent, reaffirmed its earlier judgments 596 F.2d 952 (1979). [447 U.S. 303, 307].

The Commissioner of Patents and Trademarks again sought certiorari, and we granted the writ as to both *Bergy* and *Chakrabarty*. 444 U.S. 924 (1979) Since then, *Bergy* has been dismissed as moot, 444 U.S. 1028 (1980), leaving only *Chakrabarty* for decision.

II

The Constitution grants Congress broad power to legislate to "promote the Progress of Science and useful Arts, by securing for limited Times to Authors and Inventors the exclusive Right to their respective Writings and Discoveries." Art. I, 8, cl. 8. The patent laws promote this progress by offering inventors exclusive rights for a limited period as an incentive for their inventiveness and research efforts. *Kewanee Oil Co. v. Bicron Corp.*, 416 U.S. 470, 480–481 (1974); *Universal Oil Co. v. Globe Co.*, 322 U.S. 471, 484 (1944).

Biotechnology and Genetic Engineering

The authority of Congress is exercised in the hope that "[t]he productive effort thereby fostered will have a positive effect on society through the introduction of new products and processes of manufacture into the economy, and the emanations by way of increased employment and better lives for our citizens." *Kewanee*, supra, at 480.

The question before us in this case is a narrow one of statutory interpretation requiring us to construe 35 U.S.C. 101, which provides:

"Whoever invents or discovers any new and useful process, machine, manufacture, or composition of matter, or any new and useful improvement thereof, may obtain a patent therefor, subject to the conditions and requirements of this title."

Specifically, we must determine whether respondent's micro-organism constitutes a "manufacture" or "composition of matter" within the meaning of the statute [447 U.S. 303, 308].

III

In cases of statutory construction we begin, of course, with the language of the statute. *Southeastern Community College v. Davis*, 442 U.S. 397, 405 (1979). And "unless otherwise defined, words will be interpreted as taking their ordinary, contemporary, common meaning." *Perrin v. United States*, 444 U.S. 37, 42 (1979). We have also cautioned that courts "should not read into the patent laws limitations and conditions which the legislature has not expressed." *United States v. Dubilier Condenser Corp.*, 289 U.S. 178, 199 (1933).

Guided by these canons of construction, this Court has read the term "manufacture" in 101 in accordance with its dictionary definition to mean "the production of articles for use from raw or prepared materials by giving to these materials new forms, qualities, properties, or combinations, whether by hand-labor or by machinery." *American Fruit Growers, Inc. v. Brogdex Co.*, 283 U.S. 1, 11 (1931). Similarly, "composition of matter" has been construed consistent with its common usage to include "all compositions of two or more substances and . . . all composite articles, whether they be the results of chemical union, or of mechanical mixture, or whether they be gases, fluids, powders or solids." *Shell Development Co. v. Watson*, 149 F. Supp. 279, 280 (DC 1957) (citing 1 A. Deller, *Walker on Patents* 14, p. 55 (1st ed. 1937)). In choosing such expansive terms as "manufacture" and "composition of matter," modified by the comprehensive "any," Congress plainly contemplated that the patent laws would be given wide scope.

The relevant legislative history also supports a broad construction. The Patent Act of 1793, authored by Thomas Jefferson, defined statutory subject matter as "any new and useful art, machine, manufacture, or composition of matter, or any new or useful improvement [thereof]." Act of Feb. 21, 1793, 1, 1 Stat. 319. The Act embodied Jefferson's philosophy that "ingenuity

230

should receive a liberal encouragement." [447 U.S. 303, 309] (*Writings of Thomas Jefferson* 75–76) (Washington ed. 1871). See *Graham v. John Deere Co.*, 383 U.S. 1, 7–10 (1966). Subsequent patent statutes in 1836, 1870, and 1874 employed this same broad language. In 1952, when the patent laws were recodified, Congress replaced the word "art" with "process," but otherwise left Jefferson's language intact. The Committee Reports accompanying the 1952 Act inform us that Congress intended statutory subject matter to "include anything under the sun that is made by man." S. Rep. No. 1979, 82d Cong., 2d Sess., 5 (1952); H. R. Rep. No. 1923, 82d Cong., 2d Sess., 6 (1952).

This is not to suggest that 101 has no limits or that it embraces every discovery. The laws of nature, physical phenomena, and abstract ideas have been held not patentable. See *Parker v. Flook*, 437 U.S. 584 (1978); *Gottschalk v. Benson*, 409 U.S. 63, 67 (1972); *Funk Brothers Seed Co. v. Kalo Inoculant Co.*, 333 U.S. 127, 130 (1948); *O'Reilly v. Morse*, 15 How. 62, 112–121 (1854); *Le Roy v. Tatham*, 14 How. 156, 175 (1853). Thus, a new mineral discovered in the earth or a new plant found in the wild is not patentable subject matter. Likewise, Einstein could not patent his celebrated law that $E=mc^2$.; nor could Newton have patented the law of gravity. Such discoveries are "manifestations of . . . nature, free to all men and reserved exclusively to none." *Funk*, supra, at 130.

Judged in this light, respondent's micro-organism plainly qualifies as patentable subject matter. His claim is not to a hitherto unknown natural phenomenon, but to a nonnaturally occurring manufacture or composition of matter—a product of human ingenuity "having a distinctive name, character [and] [447 U.S. 303, 310] use" *Hartranft v. Wiegmann*, 121 U.S. 609, 615 (1887). The point is underscored dramatically by comparison of the invention here with that in *Funk*. There, the patentee had discovered that there existed in nature certain species of root-nodule bacteria which did not exert a mutually inhibitive effect on each other. He used that discovery to produce a mixed culture capable of inoculating the seeds of leguminous plants. Concluding that the patentee had discovered "only some of the handiwork of nature," the Court ruled the product nonpatentable:

"Each of the species of root-nodule bacteria contained in the package infects the same group of leguminous plants which it always infected. No species acquires a different use. The combination of species produces no new bacteria, no change in the six species of bacteria, and no enlargement of the range of their utility. Each species has the same effect it always had. The bacteria perform in their natural way. Their use in combination does not improve in any way their natural functioning. They serve the ends nature originally provided and act quite independently of any effort of the patentee" [333 U.S., at 131].

Here, by contrast, the patentee has produced a new bacterium with markedly different characteristics from any found in nature and one having the potential for significant utility. His discovery is not nature's handiwork, but his own; accordingly it is patentable subject matter under 101.

IV

Two contrary arguments are advanced, neither of which we find persuasive.

(A)

The petitioner's first argument rests on the enactment of the 1930 Plant Patent Act, which afforded patent protection to certain asexually reproduced plants, and the 1970 Plant [447 U.S. 303, 311] Variety Protection Act, which authorized protection for certain sexually reproduced plants but excluded bacteria from its protection. In the petitioner's view, the passage of these Acts evidences congressional understanding that the terms "manufacture" or "composition of matter" do not include living things; if they did, the petitioner argues, neither Act would have been necessary.

We reject this argument. Prior to 1930, two factors were thought to remove plants from patent protection. The first was the belief that plants, even those artificially bred, were products of nature for purposes of the patent law. This position appears to have derived from the decision of the Patent Office in *Ex parte* Latimer, 1889 Dec. Com. Pat. 123, in which a patent claim for fiber found in the needle of the *Pinus australis* was rejected. The Commissioner reasoned that a contrary result would permit "patents [to] be obtained upon the trees of the forest and the plants of the earth, which of course would be unreasonable and impossible" Id., at 126. The Latimer case, it seems, came to "se[t] forth the general stand taken in these matters" that plants were natural products not subject to patent protection. Thorne, *Relation of Patent Law to Natural Products*, 6 J. Pat. Off. Soc. 23, 24 [447 U.S. 303, 312] (1923). The second obstacle to patent protection for plants was the fact that plants were thought not amenable to the "written description" requirement of the patent law. See 35 U.S.C. 112. Because new plants may differ from old only in color or perfume, differentiation by written description was often impossible. See Hearings on H. R. 11372 before the House Committee on Patents, 71st Cong., 2d Sess., 7 (1930) (memorandum of Patent Commissioner Robertson).

In enacting the Plant Patent Act, Congress addressed both of these concerns. It explained at length its belief that the work of the plant breeder "in aid of nature" was patentable invention. S. Rep. No. 315, 71st Cong., 2d Sess., 6–8 (1930); H. R. Rep. No. 1129, 71st Cong., 2d Sess., 7–9 (1930). And it relaxed the written description requirement in favor of "a description . . . as complete as is reasonably possible" 35 U.S.C. 162. No Committee or Member of Congress, however, expressed the broader view, now urged by the petitioner, that the terms "manufacture" or "composition of matter" exclude living things. The sole support for that position in the legislative history of the 1930 Act is found in the conclusory statement of Secretary of Agriculture Hyde, in a letter to the Chairmen of the House and Senate Committees considering the 1930 Act, that "the patent laws . . . at the present time are understood to cover only inventions or discoveries in the field of inanimate nature." See S. Rep. No. 315, supra, at Appendix A; H. R. Rep. No. 1129, supra, at Appendix A. Secretary Hyde's opinion, however, is not entitled to controlling weight. His views were solicited on the administration of the new law and not on the scope of patentable [447 U.S. 303, 313] subject matter—an area beyond his competence. Moreover, there is language in the House and Senate Committee Reports suggesting that to the extent Congress considered the matter it found the Secretary's dichotomy unpersuasive. The Reports observe:

"There is a clear and logical distinction between the discovery of a new variety of plant and of certain inanimate things, such, for example, as a new and useful natural mineral. The mineral is created wholly by nature unassisted by man. . . . On the other hand, a plant discovery resulting from cultivation is unique, isolated, and is not repeated by nature, nor can it be reproduced by nature unaided by man. . . ." S. Rep. No. 315, supra, at 6; H. R. Rep. No. 1129, supra, at 7.

Congress thus recognized that the relevant distinction was not between living and inanimate things, but between products of nature, whether living or not, and human-made inventions. Here, respondent's micro-organism is the result of human ingenuity and research. Hence, the passage of the Plant Patent Act affords the Government no support.

Nor does the passage of the 1970 Plant Variety Protection Act support the Government's position. As the Government acknowledges, sexually reproduced plants were not included under the 1930 Act because new varieties could not be reproduced true-to-type through seedlings. Brief for Petitioner 27, n. 31. By 1970, however, it was generally recognized that true-to-type reproduction was possible and that plant patent protection was therefore appropriate. The 1970 Act extended that protection. There is nothing in its language or history to suggest that it was enacted because 101 did not include living things.

In particular, we find nothing in the exclusion of bacteria from plant variety protection to support the petitioner's position. See n. 7, supra. The legislative history gives no reason for this exclusion. As the Court of Customs and [447 U.S. 303, 314] Patent Appeals suggested, it may simply reflect congressional agreement with the result reached by that court in deciding *In re Arzberger*, 27 C. C. P. A. (Pat.) 1315, 112 F.2d 834 (1940), which held that bacteria were not plants for the purposes of the 1930 Act. Or it may reflect the fact that prior to 1970 the Patent Office had issued patents for bacteria under 101. In any event, absent some clear indication that Congress "focused on [the] issues . . . directly related to the one presently before the Court," *SEC v. Sloan*, 436 U.S. 103, 120–121 (1978), there is no basis for reading into its actions an intent to modify the plain meaning of the words found in 101. See *TVA v. Hill*, 437 U.S. 153, 189–193 (1978); *United States v. Price*, 361 U.S. 304, 313 (1960).

(B)

The petitioner's second argument is that micro-organisms cannot qualify as patentable subject matter until Congress expressly authorizes such protection. His position rests on the fact that genetic technology was unforeseen when Congress enacted 101. From this it is argued that resolution of the patentability of inventions such as respondent's should be left to Congress. The legislative process, the petitioner argues, is best equipped to weigh the competing economic, social, and scientific considerations involved, and to determine whether living organisms produced by genetic engineering should receive patent protection. In support of this position, the petitioner relies on our recent holding in *Parker v. Flook*, 437 U.S. 584 (1978), and the statement that the judiciary "must proceed cautiously when . . . asked to extend [447 U.S. 303, 315] patent rights into areas wholly unforeseen by Congress." [Id., at 596].

It is, of course, correct that Congress, not the courts, must define the limits of patentability; but it is equally true that once Congress has spoken it is "the province and duty of the judicial department to say what the law is." *Marbury v. Madison*, 1 Cranch 137, 177 (1803). Congress has performed its constitutional role in defining patentable subject matter in 101; we perform ours in construing the language Congress has employed. In so doing, our obligation is to take statutes as we find them, guided, if ambiguity appears, by the legislative history and statutory purpose. Here, we perceive no ambiguity. The subject-matter provisions of the patent law have been cast in broad terms to fulfill the constitutional and statutory goal of promoting "the Progress of Science and the useful Arts" with all that means for the social and economic benefits envisioned by Jefferson. Broad general language is not necessarily ambiguous when congressional objectives require broad terms.

Nothing in *Flook* is to the contrary. That case applied our prior precedents to determine that a "claim for an improved method of calculation, even when tied to a specific end use, is unpatentable subject matter under 101." [437 U.S., at 595, n. 18]. The Court carefully scrutinized the claim at issue to determine whether it was precluded from patent protection under "the principles underlying the prohibition against patents for 'ideas' or phenomena of nature." [Id., at 593]. We have done that here. *Flook* did not announce a new principle that inventions in areas not contemplated by Congress when the patent laws were enacted are unpatentable *per se*.

Appendix A

To read that concept into *Flook* would frustrate the purposes of the patent law. This Court frequently has observed that a statute is not to be confined to the "particular application[s] . . . contemplated by the legislators." *Barr v. United States*, 324 U.S. 83, 90 (1945). Accord, *Browder v. United States*, 312 U.S. 335, 339 (1941); *Puerto Rico v. Shell Co.*, [447 U.S. 303, 316] 302 U.S. 253, 257 (1937). This is especially true in the field of patent law. A rule that unanticipated inventions are without protection would conflict with the core concept of the patent law that anticipation undermines patentability. See *Graham v. John Deere Co.*, 383 U.S., at 12–17. Mr. Justice Douglas reminded that the inventions most benefiting mankind are those that "push back the frontiers of chemistry, physics, and the like." *Great A. & P. Tea Co. v. Supermarket Corp.*, 340 U.S. 147, 154 (1950) (concurring opinion). Congress employed broad general language in drafting 101 precisely because such inventions are often unforeseeable.

To buttress his argument, the petitioner, with the support of amicus, points to grave risks that may be generated by research endeavors such as respondent's. The briefs present a gruesome parade of horribles. Scientists, among them Nobel laureates, are quoted suggesting that genetic research may pose a serious threat to the human race, or, at the very least, that the dangers are far too substantial to permit such research to proceed apace at this time. We are told that genetic research and related technological developments may spread pollution and disease, that it may result in a loss of genetic diversity, and that its practice may tend to depreciate the value of human life. These arguments are forcefully, even passionately, presented; they remind us that, at times, human ingenuity seems unable to control fully the forces it creates—that, with Hamlet, it is sometimes better "to bear those ills we have than fly to others that we know not of."

It is argued that this Court should weigh these potential hazards in considering whether respondent's invention is [447 U.S. 303, 317] patentable subject matter under 101. We disagree. The grant or denial of patents on micro-organisms is not likely to put an end to genetic research or to its attendant risks. The large amount of research that has already occurred when no researcher had sure knowledge that patent protection would be available suggests that legislative or judicial fiat as to patentability will not deter the scientific mind from probing into the unknown any more than Canute could command the tides. Whether respondent's claims are patentable may determine whether research efforts are accelerated by the hope of reward or slowed by want of incentives, but that is all.

What is more important is that we are without competence to entertain these arguments—either to brush them aside as fantasies generated by fear of the unknown, or to act on them. The choice we are urged to make is a matter of high policy for resolution within the legislative process after the kind of investigation, examination, and study that legislative bodies can provide and courts cannot. That process involves the balancing of competing values and interests, which in our democratic system is the business of elected representatives. Whatever their validity, the contentions now pressed on us should be addressed to the political branches of the Government, the Congress and the Executive, and not to the courts. [447 U.S. 303, 318].

We have emphasized in the recent past that "[o]ur individual appraisal of the wisdom or unwisdom of a particular [legislative] course . . . is to be put aside in the process of interpreting a statute." *TVA v. Hill,* 437 U.S., at 194. Our task, rather, is the narrow one of determining what Congress meant by the words it used in the statute; once that is done our powers are exhausted. Congress is free to amend 101 so as to exclude from patent protection organisms produced by genetic engineering. Cf. 42 U.S.C. 2181 (a), exempting from patent protection inventions "useful solely in the utilization of special nuclear material or atomic energy in an atomic weapon." Or it may choose to craft a statute specifically designed for such living things. But, until Congress takes such action, this Court must construe the language of 101 as it is. The language of that section fairly embraces respondent's invention.

Accordingly, the judgment of the Court of Customs and Patent Appeals is Affirmed. . . .

MR. JUSTICE BRENNAN, with whom MR. JUSTICE WHITE, MR. JUSTICE MARSHALL, and MR. JUSTICE POWELL join, dissenting.

I agree with the Court that the question before us is a narrow one. Neither the future of scientific research, nor even the ability of respondent Chakrabarty to reap some monopoly profits from his pioneering work, is at stake. Patents on the processes by which he has produced and employed the new living organism are not contested. The only question we need decide is whether Congress, exercising its authority under Art. I, 8, of the Constitution, intended that he be able to secure a monopoly on the living organism itself, no matter how produced or how used. Because I believe the Court has misread the applicable legislation, I dissent. [447 U.S. 303, 319]

The patent laws attempt to reconcile this Nation's deep-seated antipathy to monopolies with the need to encourage progress. *Deepsouth Packing Co. v. Laitram Corp.,* 406 U.S. 518, 530–531 (1972); *Graham v. John Deere Co.,* 383 U.S. 1, 7–10 (1966). Given the complexity and legislative nature of this delicate task, we must be careful to extend patent protection no further than Congress has provided. In particular, were there an absence of legislative direction, the courts should leave to Congress the decisions whether and how far to extend the patent privilege into areas where the common understanding has been that patents are not available. Cf. *Deepsouth Packing Co. v. Laitram Corp.,* supra.

Appendix A

In this case, however, we do not confront a complete legislative vacuum. The sweeping language of the Patent Act of 1793, as re-enacted in 1952, is not the last pronouncement Congress has made in this area. In 1930 Congress enacted the Plant Patent Act affording patent protection to developers of certain asexually reproduced plants. In 1970 Congress enacted the Plant Variety Protection Act to extend protection to certain new plant varieties capable of sexual reproduction. Thus, we are not dealing—as the Court would have it—with the routine problem of "unanticipated inventions." Ante, at 316. In these two Acts Congress has addressed the general problem of patenting animate inventions and has chosen carefully limited language granting protection to some kinds of discoveries, but specifically excluding others. These Acts strongly evidence a congressional limitation that excludes bacteria from patentability. [447 U.S. 303, 320]

First, the Acts evidence Congress' understanding, at least since 1930, that 101 does not include living organisms. If newly developed living organisms not naturally occurring had been patentable under 101, the plants included in the scope of the 1930 and 1970 Acts could have been patented without new legislation. Those plants, like the bacteria involved in this case, were new varieties not naturally occurring. Although the Court, ante, at 311, rejects this line of argument, it does not explain why the Acts were necessary unless to correct a pre-existing situation. I cannot share the Court's implicit assumption that Congress was engaged in either idle exercises or mere correction of the public record when it enacted the 1930 and 1970 Acts. And Congress certainly thought it was doing something significant. The Committee Reports contain expansive prose about the previously unavailable benefits to be derived from extending patent protection to plants. H. R. [447 U.S. 303, 321] Rep. No. 91-1605, pp. 1–3 (1970); S. Rep. No. 315, 71st Cong., 2d Sess., 1–3 (1930). Because Congress thought it had to legislate in order to make agricultural "human-made inventions" patentable and because the legislation Congress enacted is limited, it follows that Congress never meant to make items outside the scope of the legislation patentable.

Second, the 1970 Act clearly indicates that Congress has included bacteria within the focus of its legislative concern, but not within the scope of patent protection. Congress specifically excluded bacteria from the coverage of the 1970 Act. 7 U.S.C. 2402 (a). The Court's attempts to supply explanations for this explicit exclusion ring hollow. It is true that there is not mention in the legislative history of the exclusion, but that does not give us license to invent reasons. The fact is that Congress, assuming that animate objects as to which it had not specifically legislated could not be patented, excluded bacteria from the set of patentable organisms.

The Court protests that its holding today is dictated by the broad language of 101, which cannot "be confined to the 'particular application[s] . . .

contemplated by the legislators.'" Ante, at 315, quoting *Barr v. United States*, 324 U.S. 83, 90 (1945). But as I have shown, the Court's decision does not follow the unavoidable implications of the statute. Rather, it extends the patent system to cover living material [447 U.S. 303, 322] even though Congress plainly has legislated in the belief that 101 does not encompass living organisms. It is the role of Congress, not this Court, to broaden or narrow the reach of the patent laws. This is especially true where, as here, the composition sought to be patented uniquely implicates matters of public concern.

[footnotes omitted]

APPENDIX B

BUCK V. BELL, 274 U.S. 200 (1927)

1. The Virginia statute providing for the sexual sterilization of inmates of institutions supported by the State who shall be found to be afflicted with an hereditary form of insanity or imbecility, is within the power of the State under the Fourteenth Amendment. P. 207.

2. Failure to extend the provision to persons outside the institutions named does not render it obnoxious to the Equal Protection Clause. P. 208.

143 Va. 310, Affirmed.

Opinions

Error to a judgment of the Supreme Court of Appeals of the State of Virginia which affirmed a judgment ordering the Superintendent of the State Colony of Epileptics and Feeble Minded to perform the operation of salpingectomy on Carrie Buck, the plaintiff in error.

Holmes, J., Opinion of the Court

Mr. JUSTICE HOLMES delivered the opinion of the Court.

This is a writ of error to review a judgment of the Supreme Court of Appeals of the State of Virginia affirming a judgment of the Circuit Court of Amherst County by which the defendant in error, the superintendent of the State Colony for Epileptics and Feeble Minded, was ordered to perform the operation of salpingectomy upon Carrie Buck, the plaintiff in error, for the purpose of making her sterile. [143 Va. 310.] The case comes here upon the contention that the statute authorizing the judgment is void under the Fourteenth Amendment as denying to the plaintiff in error due process of law and the equal protection of the laws.

Carrie Buck is a feeble minded white woman who was committed to the State Colony above mentioned in due form. She is the daughter of a feeble minded mother in the same institution, and the mother of an illegitimate feeble minded child. She was eighteen years old at the time of the trial of her

239

case in the Circuit Court, in the latter part of 1924. An Act of Virginia, approved March 20, 1924, recites that the health of the patient and the welfare of society may be promoted in certain cases by the sterilization of mental defectives, under careful safeguard, &c.; that the sterilization may be effected in males by vasectomy and in females by salpingectomy, without serious pain or substantial danger to life; that the Commonwealth is supporting in various institutions many defective persons who, if now discharged, would become a menace, but, if incapable of procreating, might be discharged with safety and become self-supporting with benefit to themselves and to society, and that experience has shown that heredity plays an important part in the transmission of insanity, imbecility, &c. The statute then enacts that, whenever the superintendent of certain institutions, including the above-named State Colony, shall be of opinion that it is for the best interests of the patients and of society that an inmate under his care should be sexually sterilized, he may have the operation performed upon any patient afflicted with hereditary forms of insanity, imbecility, &c., on complying with the very careful provisions by which the act protects the patients from possible abuse.

The superintendent first presents a petition to the special board of directors of his hospital or colony, stating the facts and the grounds for his opinion, verified by affidavit. Notice of the petition and of the time and place of the hearing in the institution is to be served upon the inmate, and also upon his guardian, and if there is no guardian, the superintendent is to apply to the Circuit Court of the County to appoint one. If the inmate is a minor, notice also is to be given to his parents, if any, with a copy of the petition. The board is to see to it that the inmate may attend the hearings if desired by him or his guardian. The evidence is all to be reduced to writing, and, after the board has made its order for or against the operation, the superintendent, or the inmate, or his guardian, may appeal to the Circuit Court of the County. The Circuit Court may consider the record of the board and the evidence before it and such other admissible evidence as may be offered, and may affirm, revise, or reverse the order of the board and enter such order as it deems just. Finally any party may apply to the Supreme Court of Appeals, which, if it grants the appeal, is to hear the case upon the record of the trial in the Circuit Court, and may enter such order as it thinks the Circuit Court should have entered. There can be no doubt that, so far as procedure is concerned, the rights of the patient are most carefully considered, and, as every step in this case was taken in scrupulous compliance with the statute and after months of observation, there is no doubt that, in that respect, the plaintiff in error has had due process of law.

The attack is not upon the procedure, but upon the substantive law. It seems to be contended that in no circumstances could such an order be justified. It certainly is contended that the order cannot be justified upon the existing grounds. The judgment finds the facts that have been recited, and that Carrie Buck is the probable potential parent of socially inadequate off-spring, likewise afflicted, that she may be sexually sterilized without detriment to her general health, and that her welfare and that of society will be promoted by her sterilization, and thereupon makes the order. In view of the general declarations of the legislature and the specific findings of the Court, obviously we cannot say as matter of law that the grounds do not exist, and, if they exist, they justify the result. We have seen more than once that the public welfare may call upon the best citizens for their lives. It would be strange if it could not call upon those who already sap the strength of the State for these lesser sacrifices, often not felt to be such by those concerned, in order to prevent our being swamped with incompetence. It is better for all the world if, instead of waiting to execute degenerate offspring for crime or to let them starve for their imbecility, society can prevent those who are manifestly unfit from continuing their kind. The principle that sustains compulsory vaccination is broad enough to cover cutting the Fallopian tubes. *Jacobson v. Massachusetts*, 197 U.S. 11. Three generations of imbeciles are enough.

But, it is said, however it might be if this reasoning were applied generally, it fails when it is confined to the small number who are in the institutions named and is not applied to the multitudes outside. It is the usual last resort of constitutional arguments to point out shortcomings of this sort. But the answer is that the law does all that is needed when it does all that it can, indicates a policy, applies it to all within the lines, and seeks to bring within the lines all similarly situated so far and so fast as its means allow. Of course, so far as the operations enable those who otherwise must be kept confined to be returned to the world, and thus open the asylum to others, the equality aimed at will be more nearly reached.

APPENDIX C

NORMAN-BLOODSAW V. LAWRENCE BERKELEY LABORATORY

DOCKET 96-16526 UNITED STATES COURT OF APPEALS FOR THE NINTH CIRCUIT (1998)

MARYA S. NORMAN-BLOODSAW; EULALIO R. FUENTES; VERTIS B.ELLIS; MARK E. COVINGTON; JOHN D. RANDOLPH; ADRI-ENNE L.GARCIA; and BRENDOLYN B. SMITH, Plaintiffs-Appellants,v. No. 96-16526 LAWRENCE BERKELEY LABORATORY; CHARLES V. SHANK, Director of D.C. No. Lawrence Berkeley Laboratory; CV-95-03220-VRW HENRY H. STAUFFER, M.D.; LISA OPINIONS NOW, M.D.; T. F. BUDINGER, M.D.; WILLIAM G. DONALD, JR., M.D.; FEDERICO PENA, Secretary of the Department of Energy;* and THE REGENTS OF THE UNIVERSITY OF CALIFORNIA, a non-profit public corporation, Defendants-Appellees. Appeal from the United States District Court for the Northern District of California Vaughn R. Walker, District Judge, Presiding Argued and Submitted June 10, 1997—San Francisco, California Filed February 3, 1998

---

*Federico Pena has been substituted for his predecessor in office, Hazel O'Leary, pursuant to Fed. R. App. P. 43(c)(1). 1149

242

Appendix C

Before: Stephen Reinhardt, Thomas G. Nelson, and Michael Daly Hawkins, cCircuit Judges. Opinion by Judge Reinhardt.

This appeal involves the question whether a clerical or administrative worker who undergoes a general employee health examination may, without his knowledge, be tested for highly private and sensitive medical and genetic information such as syphilis, sickle-cell trait, and pregnancy.

Lawrence Berkeley Laboratory is a research institution jointly operated by state and federal agencies. Plaintiffs-appellants, present and former employees of Lawrence, allege that in the course of their mandatory employment entrance examinations and on subsequent occasions, Lawrence, without their knowledge or consent, tested their blood and urine for intimate medical conditions — namely, syphilis, sickle-cell trait, and pregnancy. Their complaint asserts that this testing violated Title VII of the Civil Rights Act of 1964, the Americans with Disabilities Act (ADA), and their right to privacy as guaranteed by both the United States and State of California Constitutions. The district court granted the defendants-appellees' motions for dismissal, judgment on the pleadings, and summary judgment on all of plaintiffs-appellants' claims. We affirm as to the ADA claims, but reverse as to the Title VII and state and federal privacy claims.

BACKGROUND

Plaintiffs Marya S. Norman-Bloodsaw, Eulalio R. Fuentes, Vertis B. Ellis, Mark E. Covington, John D. Randolph, Adrienne L. Garcia, and Brendolyn B. Smith are current and former administrative and clerical employees of defendant Lawrence Berkeley Laboratory ("Lawrence"), a research facility operated by the appellee Regents of the University of California pursuant to a contract with the United States Department of Energy (the Department). Defendant Charles V. Shank is the director of Lawrence, and defendants Henry H. Stauffer, Lisa Snow, T. F. Budinger, and William G. Donald, Jr., are all current or former physicians in its medical department. The named defendants are sued in both their official and individual capacities.

The Department requires federal contractors such as Lawrence to establish an occupational medical program. Since 1981, it has required its contractors to perform "preplacement examinations" of employees as part of this program, and until 1995, it also required its contractors to offer their employees the option of subsequent "periodic health examinations." The mandatory preplacement examination occurs after the offer of employment but prior to the assumption of job duties. The Department actively oversees Lawrence's occupational health program, and, prior to 1992, specifically required syphilis testing as part of the preplacement examination.

With the exception of Ellis, who was hired in 1968 and underwent an examination after beginning employment, each of the plaintiffs received written offers of employment expressly conditioned upon a "medical examination," "medical approval," or "health evaluation." All accepted these offers and underwent preplacement examinations, and Randolph and Smith underwent subsequent examinations as well.

In the course of these examinations, plaintiffs completed medical history questionnaires and provided blood and urine samples. The questionnaires asked, inter alia, whether the patient had ever had any of sixty-one medical conditions, including "[s]ickle cell anemia," "[v]enereal disease," and, in the case of women, "[m]enstrual disorders."

The blood and urine samples given by all employees during their preplacement examinations were tested for syphilis; in addition, certain samples were tested for sickle-cell trait; and certain samples were tested for pregnancy. Lawrence discontinued syphilis testing in April 1993, pregnancy testing in December 1994, and sickle-cell trait testing in June 1995. Defendants assert that they discontinued syphilis testing because of its limited usefulness in screening healthy populations, and that they discontinued sickle-cell trait testing because, by that time, most African-American adults had already been tested at birth. Lawrence continues to perform pregnancy testing, but only on an optional basis. Defendants further contend that "for many years" signs posted in the health examination rooms and "more recently" in the reception area stated that the tests at issue would be administered.

Following receipt of a right-to-sue letter from the EEOC, plaintiffs filed suit in September 1995 on behalf of all past and present Lawrence employees who have ever been subjected to the medical tests at issue. Plaintiffs allege that the testing of their blood and urine samples for syphilis, sickle-cell trait, and pregnancy occurred without their knowledge or consent, and without any subsequent notification that the tests had been conducted. They also allege that only black employees were tested for sickle-cell trait and assert the obvious fact that only female employees were tested for pregnancy.

Finally, they allege that Lawrence failed to provide safeguards to prevent the dissemination of the test results. They contend that they did not discover that the disputed tests had been conducted until approximately January 1995, and specifically deny that they observed any signs indicating that such tests would be performed. Plaintiffs do not allege that the defendants took any subsequent employment-related action on the basis of their test results, or that their test results have been disclosed to third parties.

On the basis of these factual allegations, plaintiffs contend that the defendants violated the ADA by requiring, encouraging, or assisting in medical testing that was neither job-related nor consistent with business necessity. Second, they contend that the defendants violated the federal constitutional right to privacy by conducting the testing at issue, collecting and maintaining the results of the testing, and failing to provide adequate safeguards against disclosure of the results. Third, they contend that the testing violated their right to privacy under Article I, [section] 1 of the California Constitution. Finally, plaintiffs contend that Lawrence and the Regents violated Title VII by singling out black employees for sickle-cell trait testing and by performing pregnancy testing on female employees generally.

The state defendants moved for judgment on the pleadings or, in the alternative, for summary judgment. The sole federal defendant (the "Secretary"), then-Secretary of Energy Hazel O'Leary, moved to dismiss the various claims against her for lack of subject matter jurisdiction and for failure to state a claim. Turning first to the ADA claims, the district court reasoned that because the medical questionnaires inquired into information such as venereal disease and reproductive status, plaintiffs were on notice at the time of their examinations that Lawrence was engaging in medical inquiries that were neither job-related nor consistent with business necessity. Thus, given that the most recent examination occurred over two years before the filing of the complaint, the district court held that all of the ADA claims were time-barred. It also rejected the argument that storage of the test results constitutes a "continuing violation" of the ADA that tolls the limitations period.

The district court next concluded that the federal privacy claims were also time-barred and, in the alternative, failed on the merits. On the grounds that the tests were "part of a comprehensive medical examination to which plaintiffs had consented," and that plaintiffs had completed a medical history form of "highly personal questions" that included inquiries concerning "venereal disease," "sickle-cell anemia," and "menstrual problems," it concluded that plaintiffs were aware at the time of their examinations "of sufficient facts to put them on notice" that their blood and urine would be tested for syphilis, sickle-cell trait, and pregnancy, and that their claims were thus time-barred. The district court then held, in the alternative, that the testing had not violated plaintiffs' due process right to privacy. Relying again on the fact that the tests were performed as part of a general medical examination "that covered the same areas as the tests themselves," it concluded that any "additional incremental intrusion" from the tests was so minimal that no constitutional violation could have occurred despite defendants' failure to identify "an undisputed legitimate governmental purpose" for the tests.

Finally, the district court held that the Title VII claims, even if viable, were time-barred for the same reasons as were the privacy and ADA claims. It also concluded that plaintiffs had failed to state a cognizable Title VII claim, reasoning that plaintiffs had "neither alleged nor shown any connection between these discontinued confidential tests and [their] employment terms or conditions, either in the past or in the future"; and finding that "[p]laintiffs' charge of stigmatic harm, stripped of hyperbole, speculation, and conjecture . . . evaporates."

This appeal followed.

DISCUSSION

I. STATUTE OF LIMITATIONS

[1] The district court dismissed all of the claims on statute of limitations grounds because it found that the limitations period began to run at the time the tests were taken, in which case each cause of action would be time-barred. Federal law determines when the limitations period begins to run, and the general federal rule is that "a limitations period begins to run when the plaintiff knows or has reason to know of the injury which is the basis of the action." *Trotter v. International Longshoremen's & Warehousemen's Union*, 704 F.2d 1141, 1143 (9th Cir. 1983). Because the district court resolved the statute of limitations question on summary judgment, we must determine, viewing all facts in the light most favorable to plaintiffs and resolving all factual ambiguities in their favor, whether the district court erred in determining that plaintiffs knew or should have known of the particular testing at issue when they underwent the examinations.

[2] We find that whether plaintiffs knew or had reason to know of the specific testing turns on material issues of fact that can only be resolved at trial. Plaintiffs' declarations clearly state that at the time of the examination they did not know that the testing in question would be performed, and they neither saw signs nor received any other indications to that effect. The district court had three possible reasons for concluding that plaintiffs knew or should have expected the tests at issue: (1) they submitted to an occupational preplacement examination; (2) they answered written questions as to whether they had had "venereal disease," "menstrual problems," or "sickle-cell anemia"; and (3) they voluntarily gave blood and urine samples. Given the present state of the record, these facts are hardly sufficient to establish that plaintiffs either knew or should have known that the particular testing would take place.

The question of what tests plaintiffs should have expected or foreseen depends in large part upon what preplacement medical examinations usually entail, and what, if anything, plaintiffs were told to expect. The record strongly suggests that plaintiffs' submission to the exam did not serve to afford them notice of the particular testing involved. The letters that plaintiffs received informed them merely that a "medical examination," "medical approval," or "health evaluation" was an express condition of employment. These letters did not inform plaintiffs that they would be subjected to comprehensive diagnostic medical examinations that would inquire into intimate health matters bearing no relation to their responsibilities as administrative or clerical employees.

The record, indeed, contains considerable evidence that the manner in which the tests were performed was inconsistent with sound medical practice. Plaintiffs introduced before the district court numerous expert declarations by medical scholars roundly condemning Lawrence's alleged practices and explaining, inter alia, that testing for syphilis, sickle-cell trait, and pregnancy is not an appropriate part of an occupational medical examination and is rarely if ever done by employers as a matter of routine; that Lawrence lacked any reasonable medical or public health basis for performing these tests on clerical and administrative employees such as plaintiffs; and that the performance of such tests without explicit notice and informed consent violates prevailing medical standards.

The district court also appears to have reasoned that plaintiffs knew or had reason to know of the tests because they were asked questions on a medical form concerning "venereal disease," "sickle cell anemia," and "menstrual disorders," and because they gave blood and urine samples. The fact that plaintiffs acquiesced in the minor intrusion of checking or not checking three boxes on a questionnaire does not mean that they had reason to expect further intrusions in the form of having their blood and urine tested for specific conditions that corresponded tangentially if at all to the written questions. First, the entries on the questionnaire were neither identical to nor, in some cases, even suggestive of the characteristics for which plaintiffs were tested. For example, sickle-cell trait is a genetic condition distinct from actually having sickle-cell anemia, and pregnancy is not considered a "menstrual disorder" or a "venereal disease." Second, and more important, it is not reasonable to infer that a person who answers a questionnaire upon personal knowledge is put on notice that his employer will take intrusive means to verify the accuracy of his answers. There is a significant difference between answering on the basis of what you know about your health and consenting to let someone else investigate the most intimate aspects of your life. Indeed, a reasonable person could conclude that by completing a written questionnaire, he has reduced or eliminated the need for seemingly redundant and even more intrusive laboratory testing in search of highly sensitive and non-job-related information.

Furthermore, if plaintiffs' evidence concerning reasonable medical practice is to be credited, they had no reason to think that tests would be performed without their consent simply because they had answered some questions on a form and had then, in addition, provided bodily fluid samples: Plaintiffs could reasonably have expected Lawrence to seek their consent before running any tests not usually performed in an occupational health exam—particularly tests for intimate medical conditions bearing no relationship to their responsibilities or working conditions as clerical employees. The mere fact that an employee has given a blood or urine sample does not provide notice that an employer will perform any and all tests on that specimen that it desires,—no matter how invasive—particularly where, as here, the employer has yet to offer a valid reason for the testing.

[3] In sum, the district court erred in holding as a matter of law that the plaintiffs knew or had reason to know of the nature of the tests as a result of their submission to the preemployment medical examinations. Because the question of what testing, if any, plaintiffs had reason to expect turns on material factual issues that can only be resolved at trial, summary judgment on statute of limitations grounds was inappropriate with respect to the causes of action based on an invasion of privacy in violation of the Federal and California Constitutions, and also on the Title VII claims.

II. FEDERAL CONSTITUTIONAL DUE PROCESS RIGHT OF PRIVACY

The district court also ruled, in the alternative, on the merits of all of plaintiffs' claims except the ADA claims. We first examine its ruling with respect to the claim for violation of the federal constitutional right to privacy. While acknowledging that the government had failed to identify any "undisputed legitimate governmental purpose" for the three tests, the district court concluded that no violation of plaintiffs' right to privacy could have occurred because any intrusions arising from the testing were *de minimis* in light of (1) the "large overlap" between the subjects covered by the medical questionnaire and the three tests and (2) the "overall intrusiveness" of "a full-scale physical examination." We hold that the district court erred.

Because the ADA claims fail on the merits, as discussed below, we do not determine whether the district court erred in dismissing those claims on statute of limitations grounds.

[4] The constitutionally protected privacy interest in avoiding disclosure of personal matters clearly encompasses medical information and its confidentiality. *Doe v. Attorney General of the United States*, 941 F.2d 780, 795 (9th Cir. 1991) (citing *United States v. Westinghouse Elec. Corp.*, 638 F.2d 570, 577 (3d Cir. 1980)); *Roe v. Sherry*, 91 F.3d 1270, 1274 (9th Cir. 1996); see also *Doe v. City of New York*, 15 F.3d 264, 267–69 (2d Cir. 1994). Although cases defining the privacy interest in medical information have typically involved its disclosure to "third" parties, rather than the collection of information by illicit

means, it goes without saying that the most basic violation possible involves the performance of unauthorized tests—that is, the non-consensual retrieval of previously unrevealed medical information that may be unknown even to plaintiffs. These tests may also be viewed as searches in violation of Fourth Amendment rights that require Fourth Amendment scrutiny. The tests at issue in this case thus implicate rights protected under both the Fourth Amendment and the Due Process Clause of the Fifth or Fourteenth Amendments. *Yin v. California*, 95 F.3d 864, 870 (9th Cir. 1996), cert. denied, 117 S. Ct. 955 (1997).

[5] Because it would not make sense to examine the collection of medical information under two different approaches, we generally "analyze [medical tests and examinations] under the rubric of [the Fourth] Amendment. " Id. at 871 & n.12. Accordingly, we must balance the government's interest in conducting these particular tests against the plaintiffs' expectations of privacy. Id. at 873. Furthermore, "application of the balancing test requires not only considering the degree of intrusiveness and the state's interests in requiring that intrusion, but also 'the efficacy of this [the state's] means for meeting' its needs." Id. (quoting *Vernonia Sch. Dist. 47J v. Acton*, 515 U.S. 646, 660 (1995)).

[6] The district court erred in dismissing the claims on the ground that any violation was *de minimis*, incremental, or overlapping. The latter two grounds are actually just the court's explanations for its adoption of its "*de minimis*" conclusion. They are not in themselves reasons for dismissal. Nor if the violation is otherwise significant does it become insignificant simply because it is overlapping or incremental. We cannot, therefore, escape a scrupulous examination of the nature of the violation, although we can, of course, consider whether the plaintiffs have in fact consented to any part of the alleged intrusion.

[7] One can think of few subject areas more personal and more likely to implicate privacy interests than that of one's health or genetic make-up. Doe, 15 F.3d at 267 ("Extension of the right to confidentiality to personal medical information recognizes there are few matters that are quite so personal as the status of one's health"); see *Vernonia Sch. Dist. 47J*, 515 U.S. at 658 (noting under Fourth Amendment analysis that "it is significant that the tests at issue here look only for drugs, and not for whether the student is, for example, epileptic, pregnant, or diabetic"). Furthermore, the facts revealed by the tests are highly sensitive, even relative to other medical information. With respect to the testing of plaintiffs for syphilis and pregnancy, it is well established in this circuit "that the Constitution prohibits unregulated, unrestrained employer inquiries into personal sexual matters that have no bearing on job performance." *Schowengerdt v. General Dynamics Corp.*, 823 F.2d 1328, 1336 (9th Cir. 1987) (citing *Thorne v. City of El Segundo*, 726 F.2d 459, 470 (9th Cir.

1983)). The fact that one has syphilis is an intimate matter that pertains to one's sexual history and may invite tremendous amounts of social stigma. Pregnancy is likewise, for many, an intensely private matter, which also may pertain to one's sexual history and often carries far-reaching societal implications. See *Thorne*, 726 F.2d at 468–70; Doe, 15 F.3d at 267 (noting discrimination and intolerance to which HIV-positive persons are exposed). Finally, the carrying of sickle-cell trait can pertain to sensitive information about family history and reproductive decisionmaking. Thus, the conditions tested for were aspects of one's health in which one enjoys the highest expectations of privacy.

[8] As discussed above, with respect to the question of the statute of limitations, there was little, if any, "overlap" between what plaintiffs consented to and the testing at issue here. Nor was the additional invasion only incremental. In some instances, the tests related to entirely different conditions. In all, the information obtained as the result of the testing was qualitatively different from the information that plaintiffs provided in their answers to the questions, and was highly invasive. That one has consented to a general medical examination does not abolish one's privacy right not to be tested for intimate, personal matters involving one's health—nor does consenting to giving blood or urine samples, or filling out a questionnaire. As we have made clear, revealing one's personal knowledge as to whether one has a particular medical condition has nothing to do with one's expectations about actually being tested for that condition. Thus, the intrusion was by no means de minimis. Rather, if unauthorized, the testing constituted a significant invasion of a right that is of great importance, and labelling it minimal cannot and does not make it so.

[9] Lawrence further contends that the tests in question, even if their intrusiveness is not de minimis, would be justified by an employer's interest in performing a general physical examination. This argument fails because issues of fact exist with respect to whether the testing at issue is normally part of a general physical examination. There would of course be no violation if the testing were authorized, or if the plaintiffs reasonably should have known that the blood and urine samples they provided would be used for the disputed testing and failed to object. However, as we concluded in Section I, material issues of fact exist as to those questions.

Summary judgment in the alternative on the merits of the federal constitutional privacy claim was therefore incorrect.

III. RIGHT TO PRIVACY UNDER ARTICLE I, [SECTION] 1 OF THE CALIFORNIA CONSTITUTION

With respect to the state privacy claims, defendants argue, as they did with respect to the federal privacy claims, that the intrusions occasioned by the testing were so minimal that the government need not demonstrate a legitimate interest in performing the tests. In the alternative, they argue that the intrusions were so minimal that plaintiffs' privacy interests were necessarily overcome by the government's interest in performing the preplacement examinations. We understand this argument to be essentially the same as the argument that these tests are a part of an ordinary general medical examination. Defendants urge no additional governmental interest but appear to rely entirely on the interest that any employer might assert in requiring potential employees to undergo general medical testing. The district court did not adopt either of the defendants' positions expressly but simply ruled that plaintiffs "could not proceed" because the "undisputed facts"—namely, completion of the medical questionnaire, consent to the preplacement examination, and the voluntary giving of blood and urine samples—showed that the tests had inflicted "only a *de minimis* privacy invasion."

[10] To assert a cause of action under Article I, S 1 of the California Constitution, one must establish three elements: (1) a legally protected privacy interest; (2) a reasonable expectation of privacy under the circumstances; and (3) conduct by the defendant that amounts to a "serious invasion" of the protected privacy interest. *Loder v. City of Glendale*, 927 P.2d 1200, 1228 (Cal. 1997) (quoting *Hill v. National Collegiate Athletic Ass'n*, 865 P.2d 633, 657 (Cal. 1994), cert. denied, 118 S.Ct. 44 (1997)). These elements must be "viewed simply as 'threshold elements,'" after which the court must conduct a balancing test between the "countervailing interests" for the conduct in question and the intrusion on privacy resulting from the conduct. A showing of "countervailing interests" may, in turn, be rebutted by a showing that there were "feasible and effective alternatives" with a "lesser impact on privacy interests." *Hill*, 865 P.2d at 657.

[11] For much the same reasons as we have discussed above with respect to the statute of limitations and federal privacy claims, the district court erred in dismissing the state constitutional privacy claim. The only possible difference between the state claim and the federal claim is the threshold requirement that the invasion be serious, and for purposes of summary judgment, that requirement has been more than met.

For the reasons discussed above, we find that material issues of fact exist with respect to whether the defendants had any interest at all in obtaining the information and whether plaintiffs had a reasonable expectation of privacy under the circumstances. Both these questions involve a factual dispute regarding the ordinary or accepted medical practice regarding general or pre-employment medical exams.

Accordingly, the district court also erred in dismissing the state constitutional privacy claims.

IV. TITLE VII CLAIMS

The district court also dismissed the Title VII counts on the merits on the grounds that plaintiffs had failed to state a claim because the "alleged classifications, standing alone, do not suffice to provide a cognizable basis for relief under Title VII" and because plaintiffs had neither alleged nor demonstrated how these classifications had adversely affected them.

[12] Section 703(a) of Title VII of the Civil Rights Act of 1964 provides that it is unlawful for any employer:

(1) to fail or refuse to hire or to discharge any individual, or otherwise to discriminate against any individual with respect to his compensation, terms, conditions, or privileges of employment, because of such individual's race, color, religion, sex, or national origin; or

(2) to limit, segregate, or classify his employees or applicants for employment in any way which would deprive or tend to deprive any individual of employment opportunities or otherwise adversely affect his status as an employee, because of such individual's race, color, religion, sex, or national origin.

42 U.S.C. S 2000e-2(a)

The Pregnancy Discrimination Act further provides that discrimination on the basis of "sex" includes discrimination "on the basis of pregnancy, childbirth, or related medical conditions." 42 U.S.C. S 2000e(k). "In accordance with Congressional intent, the above language is to be read in the broadest possible terms. The intent of Congress was not to list specific discriminatory practices, nor to definitively set out the scope of the activities covered." EEOC Compliance Manual (CCH) S 613.1, at P 2901 (citing *Rogers v. EEOC*, 454 F.2d 234 (5th Cir. 1971)).

Despite defendants' assertions to the contrary, plaintiffs' Title VII claims fall neatly into a Title VII framework: Plaintiffs allege that black and female employees were singled out for additional nonconsensual testing and that defendants thus selectively invaded the privacy of certain employees on the basis of race, sex, and pregnancy. The district court held that

(1) the tests did not constitute discrimination in the "terms" or "conditions" of plaintiffs' employment; and that (2) plaintiffs have failed to show any "adverse effect" as a result of the tests. It also granted the plaintiffs leave to amend their complaint to show adverse effect.

[13] Under [section] 2000e-2(a)(1), supra, an employer who "otherwise . . . discriminate[s]" with respect to the "terms" or "conditions" of employment on account of an illicit classification is subject to Title VII liability. It is well established that Title VII bars discrimination not only in the "terms" and "conditions" of ongoing employment, but also in the "terms" and "conditions" under which individuals may obtain employment. See, e.g., *Griggs v.*

Appendix C

Duke Power Co., 401 U.S. 424, 432–37 (1971) (facially neutral educational and testing requirements that are not reasonable measures of job performance and have disparate impact on hiring of minorities violate Title VII). Thus, for example, a requirement of preemployment health examinations imposed only on female employees, or a requirement of preemployment background security checks imposed only on black employees, would surely violate Title VII.

[14] In this case, the term or condition for black employees was undergoing a test for sickle-cell trait; for women it was undergoing a test for pregnancy. It is not disputed that the preplacement exams were, literally, a condition of employment: the offers of employment stated this explicitly. Thus, the employment of women and blacks at Lawrence was conditioned in part on allegedly unconstitutional invasions of privacy to which white and/or male employees were not subjected. An additional "term or condition" requiring an unconstitutional invasion of privacy is, without doubt, actionable under Title VII. Furthermore, even if the intrusions did not rise to the level of unconstitutionality, they would still be a "term" or "condition" based on an illicit category as described by the statute and thus a proper basis for a Title VII action. Thus, the district court erred in ruling on the leadings that the plaintiffs had failed to assert a proper Title VII claim under [section] 2000e-2(a)(1).

[15] The district court also erred in finding as a matter of law that there was no "adverse effect" with respect to the tests as required under [section] 2000e-2(a)(2). The unauthorized obtaining of sensitive medical information on the basis of race or sex would in itself constitute an "adverse effect," or injury, under Title VII.

Thus, it was error to rule that as a matter of law no "adverse effect" could arise from a classification that singled out particular groups for unconstitutionally invasive, non-consensual medical testing, and the district court erred in dismissing the Title VII claims on this ground as well.

V. THE ADA CLAIMS

Plaintiffs may challenge only the medical examinations that occurred "on or after January 26, 1992," which is the effective date of the ADA for public entities. The only plaintiffs who underwent any examinations or testing on or after that date are Fuentes and Garcia, who were tested in April 1992 and August 1993, respectively. The complaint alleges that defendants violated the ADA by requiring medical examinations and making medical inquiries that were "neither job-related nor consistent with business necessity." (Compl.P 64 (citing 42 U.S.C. [section] 12112(c)(4)). On appeal, plaintiffs also argue that "the ADA limits medical record keeping by an employer to the results of job-related examinations consistent with business necessity." Appellant Br. at 49 (citing 42 U.S.C.[section] 12112(d)). Plaintiffs do not allege that defendants made use of information gathered in the examinations to discriminate against them on the basis of disability; indeed, neither Garcia nor Fuentes received any positive test results.

[16] The ADA creates three categories of medical inquiries and examinations by employers: (1) those conducted prior to an offer of employment ("preemployment" inquiries and examinations); (2) those conducted "after an offer of employment has been made" but "prior to the commencement of . . . employment duties" ("employment entrance examinations"); and (3) those conducted at any point thereafter. It is undisputed that the second category, employment entrance examinations, as governed by [section] 12112(d)(3), are the examinations and inquiries to which Fuentes and Garcia were subjected. Unlike examinations conducted at any other time, an employment entrance examination need not be concerned solely with the individual's "ability to perform job-related functions," [section] 12112(d)(2); nor must it be "job-related or consistent with business necessity," [section] 12112(d)(4). Thus, the ADA imposes no restriction on the scope of entrance examinations; it only guarantees the confidentiality of the information gathered, [section] 12112(d)(3)(B), and restricts the use to which an employer may put the information. [section] 12112(d)(3)(C); see 42 U.S.C.[section] 12112(d)(1) (medical examinations and inquiries must be consistent with the general prohibition in [section] 12112(a) against discrimination on the basis of disability); 29 C.F.R. [section] 1630.14(b)(3) (if the results of the examination exclude an individual on the basis of disability, the exclusionary criteria themselves must be job-related and consistent with business necessity). Because the ADA does not limit the scope of such examinations to matters that are "job-related and consistent with business necessity," dismissal of the ADA claims was proper.

Plaintiffs' new argument on appeal that the ADA limits medical record-keeping to "the results of job-related examinations consistent with business necessity" also lacks merit. Section 12112(d)(3)(B) sets forth the conditions under which information obtained during the entrance examination must be kept but clearly does not purport to restrict the records that may be kept to matters that are "job-related and consistent with business necessity." Thus plaintiffs' ADA claims also fail in this respect.

The only possible ADA claim is directed at the defendants' alleged failure to maintain plaintiffs' medical records in the manner required by [section] 12112(d)(3)(B). The allegations in plaintiffs' complaint do not explicitly set forth such a violation but incorporate by reference the factual allegation that the defendants "[f]ail[ed] to provide safeguards to prevent the dissemination to third parties of sensitive medical information regarding the plaintiffs." On appeal the plaintiffs argue only that the defendants have "failed to describe the procedures by which a third party might gain access to the records, and the enforcement of any rules, policies, regulations or procedures to prevent third parties from gaining access to the records." To the extent that one can construe the complaint to allege that the defendants are in violation of [section] 12112(d)(3)(B), the bare allegation that defendants have not provided, or adequately described, safeguards fails to state a violation of the ADA requirements as set forth in [section] 12112(d)(3)(B) or as implemented in Department orders. See DOE Order 440.1 (Sep. 30, 1995); DOE Order 5480.8A (June 6, 1992); DOE Order 5480.8 (May 22, 1981). Accordingly, dismissal of the ADA claims was proper.

VI. PLAINTIFFS' CLAIMS ARE NOT MOOT

The Secretary contends that the claims against him in his official capacity for injunctive and declaratory relief are moot because (1) the only testing that the Department ever required was syphilis testing, and (2) the DOE order that required syphilis testing was cancelled on June 22, 1992, and replaced by a different order that requires "[u]rinalysis and serology" only "when indicated." Compare DOE Order 5480.8 (May 22, 1981), with DOE Order 5480.8A (June 26, 1992).

Although the state defendants do not raise the issue, a similar argument can be made on their behalf: Lawrence discontinued syphilis testing in April 1993, pregnancy testing in December 1994, and sickle-cell trait testing in June 1995.

[17] "[A] case is moot when the issues presented are no longer 'live' or the parties lack a legally cognizable interest in the outcome." *County of Los Angeles v. Davis*, 440 U.S.625, 631 (1979) [quoting *Powell v. McCormack*, 395 U.S. 486, 496 (1969)]. "Mere voluntary cessation of allegedly illegal conduct does not moot a case; it if did, the courts would be compelled to leave [t]he defendant . . . free to return to his old ways." *United States v. Concentrated Phosphate Export Ass'n*, 393 U.S. 199, 203 (1968) (quoting *United States v. W. T. Grant Co.*, 345 U.S. 629, 632 (1953)). Nevertheless, part or all of a case may become moot if (1) "subsequent events [have] made it absolutely clear that the allegedly wrongful behavior [cannot] reasonably be expected to recur," *Concentrated Phosphate*, 393 U.S. at 203, and (2) "interim relief or events have completely and irrevocably eradicated the effects of the alleged violation." *Lindquist v. Idaho State Bd. of Corrections*, 776 F.2d 851, 854 (9th Cir. 1985) (quoting *Davis*, 440 U.S. at 631). "The burden of demonstrating mootness 'is a heavy one.'" *Davis*, 440 U.S. at 631 (quoting *W. T. Grant*, 345 U.S. at 632–33).

[18] Defendants have not carried their heavy burden of establishing either that their alleged behavior cannot be reasonably expected to recur, or that

interim events have eradicated the effects of the alleged violation. First, they do not contend that the Department will never again require or permit, or that Lawrence will never again conduct, the tests at issue. They assert only that syphilis testing was discontinued because of its limited usefulness in screening healthy populations, and that sickle-cell trait testing was discontinued as redundant of testing that most African Americans now receive at birth. Moreover, in the case of pregnancy testing, they do not even argue that such testing is no longer medically useful; rather, they have simply made it optional. Defendants have neither asserted nor demonstrated that they will never resume mandatory testing for intimate medical conditions; nor have they offered any reason why they might not return in the future to their original views on the utility of mandatory testing. In contrast, plaintiffs have introduced evidence, in the form of correspondence between Lawrence and the department, that the syphilis tests were discontinued merely for reasons of "cost-effectiveness." See *Concentrated Phosphate*, 393 U.S. at 203 (holding that mere statement that it would be "uneconomical" for defendants to continue their allegedly wrongful conduct "cannot suffice to satisfy the heavy burden" of establishing mootness).

[19] Second, defendants also have not asserted that any "interim relief or events have completely and irrevocably eradicated the effects of the alleged violation." *Lindquist*, 776 F.2d at 854. Indeed, it is undisputed that the Department requires Lawrence to retain plaintiffs' test results and that Lawrence does in fact do so. See DOE Order 440.1, dated September 30, 1995 ("Employee medical records shall be adequately protected and stored permanently.") Even if the continued storage, against plaintiffs' wishes, of intimate medical information that was allegedly taken from them by unconstitutional means does not itself constitute a violation of law, it is clearly an ongoing "effect" of the allegedly unconstitutional and discriminatory testing, and expungement of the test results would be an appropriate remedy for the alleged violation. Cf. *Fendler v. United States Parole Comm'n*, 774 F.2d 975, 979 (9th Cir. 1985) ("Federal courts have the equitable power 'to order the expungement of Government records where necessary to vindicate rights secured by the Constitution or by statute.") [quoting *Chastain v. Kelley*, 510 F.2d 1232, 1235 (D.C. Cir. 1975)]; *Maurer v. Pitchess*, 691 F.2d 434, 437 (9th Cir. 1982). Accordingly, plaintiffs' claims for injunctive and declaratory relief are not moot.

VII. IRREPARABLE INJURY

[20] Finally, the Secretary contends that plaintiffs cannot seek injunctive relief because they have not alleged irreparable injury. To obtain injunctive relief, " '[a] reasonable showing' of a 'sufficient likelihood' that plaintiff will

be injured again is necessary." *Kruse v. State of Hawaii*, 68 F.3d 331, 335 (9th Cir. 1995) (internal quotation marks omitted); see *City of Los Angeles v. Lyons*, 461 U.S. 95, 111 (1983). "The likelihood of the injury recurring must be calculable and if there is no basis for predicting that any future repetition would affect the present plaintiffs, there is no case or controversy." *Sample v. Johnson*, 771 F.2d 1335, 1340 (9th Cir. 1985). In this case, plaintiffs seek not only to enjoin future illegal testing, but also to require defendants, inter alia, to notify all employees who may have been tested illegally; to destroy the results of such illegal testing upon employee request; to describe any use to which the information was put, and any disclosures of the information that were made; and to submit Lawrence's medical department to "independent oversight and monitoring."

[21] At the very least, the retention of undisputedly intimate medical information obtained in an unconstitutional and discriminatory manner would constitute a continuing "irreparable injury" for purposes of equitable relief. Moreover, the Department orders still require Lawrence to conduct preplacement examinations. DOE Order 440.1 (Sep. 30, 1995). Thus, there seems to be at least a reasonable possibility that Lawrence would again conduct undisclosed medical testing of its employees for intimate medical conditions. For these reasons, a request for injunctive relief is proper.

CONCLUSION

Because material and disputed issues of fact exist with respect to whether reasonable persons in plaintiffs' position would have had reason to know that the tests were being performed, and because the tests were a separate and more invasive intrusion into their privacy than the aspects of the examination to which they did consent, the district court erred in granting summary judgment on statute of limitations grounds with respect to the Title VII claims and the federal and state constitutional privacy claims. The district court also erred in dismissing the federal and state constitutional privacy claims and the Title VII claims on the merits. The district court's dismissal of the ADA claims was proper. None of the Secretary's arguments with respect to the claims brought against him in his official capacity has merit.

AFFIRMED IN PART, REVERSED IN PART, AND REMANDED.

[footnotes omitted]

APPENDIX D

BRAGDON V. ABBOTT
97 U.S. 156 (1998)

[portions are omitted]

ON WRIT OF CERTIORARI TO THE UNITED STATES COURT OF APPEALS FOR THE FIRST CIRCUIT

Justice Kennedy delivered the opinion of the Court.

We address in this case the application of the Americans with Disabilities Act of 1990 (ADA), 104 Stat. 327, 42 U.S.C. § 12101 et seq., to persons infected with the human immunodeficiency virus (HIV). We granted certiorari to review, first, whether HIV infection is a disability under the ADA when the infection has not yet progressed to the so-called symptomatic phase; and, second, whether the Court of Appeals, in affirming a grant of summary judgment, cited sufficient material in the record to determine, as a matter of law, that respondent's infection with HIV posed no direct threat to the health and safety of her treating dentist.

I

Respondent Sidney Abbott has been infected with HIV since 1986. When the incidents we recite occurred, her infection had not manifested its most serious symptoms. On September 16, 1994, she went to the office of petitioner Randon Bragdon in Bangor, Maine, for a dental appointment. She disclosed her HIV infection on the patient registration form. Petitioner completed a dental examination, discovered a cavity, and informed respondent of his policy against filling cavities of HIV-infected patients. He offered to perform the work at a hospital with no added fee for his services, though respondent would be responsible for the cost of using the hospital's facilities. Respondent declined.

Respondent sued petitioner under state law and §302 of the ADA, 104 Stat. 355, 42 U.S.C. § 12182 alleging discrimination on the basis of her disability. The state law claims are not before us. Section 302 of the ADA provides:

"No individual shall be discriminated against on the basis of disability in the full and equal enjoyment of the goods, services, facilities, privileges, advantages, or accommodations of any place of public accommodation by any person who . . . operates a place of public accommodation." §12182(a).

The term "public accommodation" is defined to include the "professional office of a health care provider." §12181(7)(F).

A later subsection qualifies the mandate not to discriminate. It provides:

"Nothing in this subchapter shall require an entity to permit an individual to participate in or benefit from the goods, services, facilities, privileges, advantages and accommodations of such entity where such individual poses a direct threat to the health or safety of others." §12182(b)(3).

The United States and the Maine Human Rights Commission intervened as plaintiffs. After discovery, the parties filed cross-motions for summary judgment. The District Court ruled in favor of the plaintiffs, holding that respondent's HIV infection satisfied the ADA's definition of disability. 912 F. Supp. 580, 585–587 (Me. 1995). The court held further that petitioner raised no genuine issue of material fact as to whether respondent's HIV infection would have posed a direct threat to the health or safety of others during the course of a dental treatment. Id., at 587–591. The court relied on affidavits submitted by Dr. Donald Wayne Marianos, Director of the Division of Oral Health of the Centers for Disease Control and Prevention (CDC). The Marianos affidavits asserted it is safe for dentists to treat patients infected with HIV in dental offices if the dentist follows the so-called universal precautions described in the *Recommended Infection-Control Practices for Dentistry* issued by CDC in 1993 (1993 CDC *Dentistry Guidelines*). 912 F. Supp., at 589.

The Court of Appeals affirmed. It held respondent's HIV infection was a disability under the ADA, even though her infection had not yet progressed to the symptomatic stage. 107 F.3d 934, 939–943 (CA1 1997). The Court of Appeals also agreed that treating the respondent in petitioner's office would not have posed a direct threat to the health and safety of others. Id., at 943–948. Unlike the District Court, however, the Court of Appeals declined to rely on the Marianos affidavits. Id., at 946, n. 7. Instead the court relied on the 1993 CDC *Dentistry Guidelines*, as well as the *Policy on AIDS, HIV Infection and the Practice of Dentistry*, promulgated by the American Dental Association in 1991 (1991 American Dental Association *Policy on HIV*). 107 F.3d, at 945–946.

II

We first review the ruling that respondent's HIV infection constituted a disability under the ADA. The statute defines disability as:

"(A) a physical or mental impairment that substantially limits one or more of the major life activities of such individual;

"(B) a record of such an impairment; or

"(C) being regarded as having such impairment." §12102(2).

We hold respondent's HIV infection was a disability under subsection (A) of the definitional section of the statute. In light of this conclusion, we need not consider the applicability of subsections (B) or (C).

Our consideration of subsection (A) of the definition proceeds in three steps. First, we consider whether respondent's HIV infection was a physical impairment. Second, we identify the life activity upon which respondent relies (reproduction and child bearing) and determine whether it constitutes a major life activity under the ADA. Third, tying the two statutory phrases together, we ask whether the impairment substantially limited the major life activity. In construing the statute, we are informed by interpretations of parallel definitions in previous statutes and the views of various administrative agencies which have faced this interpretive question.

A

The ADA's definition of disability is drawn almost verbatim from the definition of "handicapped individual" included in the Rehabilitation Act of 1973, 29 U.S.C. § 706(8)(B) (1988 ed.), and the definition of "handicap" contained in the Fair Housing Amendments Act of 1988, 42 U.S.C. § 3602(h)(1) (1988 ed.). Congress' repetition of a well-established term carries the implication that Congress intended the term to be construed in accordance with pre-existing regulatory interpretations. See *FDIC v. Philadelphia Gear Corp.*, 476 U.S. 426, 437–438 (1986); *Commissioner v. Estate of Noel*, 380 U.S. 678, 681–682 (1965); *ICC v. Parker*, 326 U.S. 60, 65 (1945). In this case, Congress did more than suggest this construction; it adopted a specific statutory provision in the ADA directing as follows:

"Except as otherwise provided in this chapter, nothing in this chapter shall be construed to apply a lesser standard than the standards applied under title V of the Rehabilitation Act of 1973 (29 U.S.C. 790 *et seq*.) or the regulations issued by Federal agencies pursuant to such title." 42 U.S.C. § 12201(a).

The directive requires us to construe the ADA to grant at least as much protection as provided by the regulations implementing the Rehabilitation Act.

1. The first step in the inquiry under subsection (A) requires us to determine whether respondent's condition constituted a physical impairment. The Department of Health, Education and Welfare (HEW) issued the first regulations interpreting the Rehabilitation Act in 1977. The regulations are of particular significance because, at the time, HEW was the agency responsible for coordinating the implementation and enforcement of §504. *Consolidated Rail Corporation v. Darrone*, 465 U.S. 624, 634, (1984) [citing Exec. Order No. 11914, 3 CFR 117 (1976–1980 Comp.)]. The HEW regulations, which appear without change in the current regulations issued by the

Department of Health and Human Services, define "physical or mental impairment" to mean:

"(A) any physiological disorder or condition, cosmetic disfigurement, or anatomical loss affecting one or more of the following body systems: neurological; musculoskeletal; special sense organs; respiratory, including speech organs; cardiovascular; reproductive, digestive, genito-urinary; hemic and lymphatic; skin; and endocrine; or

"(B) any mental or psychological disorder, such as mental retardation, organic brain syndrome, emotional or mental illness, and specific learning disabilities." 45 CFR § 84.3(j)(2)(i) (1997).

In issuing these regulations, HEW decided against including a list of disorders constituting physical or mental impairments, out of concern that any specific enumeration might not be comprehensive. . . . [material establishing that asymptomatic HIV infection is a disability is omitted]

In light of the immediacy with which the virus begins to damage the infected person's white blood cells and the severity of the disease, we hold it is an impairment from the moment of infection. As noted earlier, infection with HIV causes immediate abnormalities in a person's blood, and the infected person's white cell count continues to drop throughout the course of the disease, even when the attack is concentrated in the lymph nodes. In light of these facts, HIV infection must be regarded as a physiological disorder with a constant and detrimental effect on the infected person's hemic and lymphatic systems from the moment of infection. HIV infection satisfies the statutory and regulatory definition of a physical impairment during every stage of the disease.

2. The statute is not operative, and the definition not satisfied, unless the impairment affects a major life activity. Respondent's claim throughout this case has been that the HIV infection placed a substantial limitation on her ability to reproduce and to bear children. App. 14; 912 F. Supp., at 586; 107 F.3d, at 939. Given the pervasive, and invariably fatal, course of the disease, its effect on major life activities of many sorts might have been relevant to our inquiry. Respondent and a number of amici make arguments about HIV's profound impact on almost every phase of the infected person's life. See Brief for Respondent Sidney Abbott 24–27; Brief for American Medical Association as *Amicus Curiae* 20; Brief for Infectious Diseases Society of America et al. as *Amici Curiae* 7–11. In light of these submissions, it may seem legalistic to circumscribe our discussion to the activity of reproduction. We have little doubt that had different parties brought the suit they would have maintained that an HIV infection imposes substantial limitations on other major life activities.

From the outset, however, the case has been treated as one in which reproduction was the major life activity limited by the impairment. It is our practice to decide cases on the grounds raised and considered in the Court of Appeals and included in the question on which we granted certiorari. See, e.g., *Blessing v. Freestone*, 520 U.S. 329, 340, n. 3 (1997) (citing this Court's Rule 14.1(a)); *Capitol Square Review and Advisory Bd. v. Pinette*, 515 U.S. 753, 760 (1995). We ask, then, whether reproduction is a major life activity.

We have little difficulty concluding that it is. As the Court of Appeals held, "[t]he plain meaning of the word 'major' denotes comparative importance" and "suggest[s] that the touchstone for determining an activity's inclusion under the statutory rubric is its significance." 107 F.3d, at 939, 940. Reproduction falls well within the phrase "major life activity." Reproduction and the sexual dynamics surrounding it are central to the life process itself.

While petitioner concedes the importance of reproduction, he claims that Congress intended the ADA only to cover those aspects of a person's life which have a public, economic, or daily character. Brief for Petitioner 14, 28, 30, 31; see also id., at 36–37 [citing *Krauel v. Iowa Methodist Medical Center*, 95 F.3d 674, 677 (CA8 1996)]. The argument founders on the statutory language. Nothing in the definition suggests that activities without a public, economic, or daily dimension may somehow be regarded as so unimportant or insignificant as to fall outside the meaning of the word "major." The breadth of the term confounds the attempt to limit its construction in this manner.

As we have noted, the ADA must be construed to be consistent with regulations issued to implement the Rehabilitation Act. See 42 U.S.C. § 12201(a). Rather than enunciating a general principle for determining what is and is not a major life activity, the Rehabilitation Act regulations instead provide a representative list, defining term to include "functions such as caring for one's self, performing manual tasks, walking, seeing, hearing, speaking, breathing, learning, and working." 45 CFR § 84.3(j)(2)(ii) (1997); 28 CFR § 41.31(b)(2) (1997). As the use of the term "such as" confirms, the list is illustrative, not exhaustive.

These regulations are contrary to petitioner's attempt to limit the meaning of the term "major" to public activities. The inclusion of activities such as caring for one's self and performing manual tasks belies the suggestion that a task must have a public or economic character in order to be a major life activity for purposes of the ADA. On the contrary, the Rehabilitation Act regulations support the inclusion of reproduction as a major life activity, since reproduction could not be regarded as any less important than working and learning. Petitioner advances no credible basis for confining major life activities to those with a public, economic, or daily aspect. In the absence of any reason to reach a contrary conclusion, we agree with the Court of Appeals' determination that reproduction is a major life activity for the purposes of the ADA.

Appendix D

3. The final element of the disability definition in subsection (A) is whether respondent's physical impairment was a substantial limit on the major life activity she asserts. The Rehabilitation Act regulations provide no additional guidance. 45 CFR pt. 84, App. A, p. 334 (1997).

Our evaluation of the medical evidence leads us to conclude that respondent's infection substantially limited her ability to reproduce in two independent ways. First, a woman infected with HIV who tries to conceive a child imposes on the man a significant risk of becoming infected. The cumulative results of 13 studies collected in a 1994 textbook on AIDS indicates that 20% of male partners of women with HIV became HIV-positive themselves, with a majority of the studies finding a statistically significant risk of infection. Osmond & Padian, "Sexual Transmission of HIV," in *AIDS Knowledge Base* 1.9-8, and tbl. 2; see also Haverkos & Battjes, "Female-to-Male Transmission of HIV," 268 *JAMA* 1855, 1856, tbl. (1992) (cumulative results of 16 studies indicated 25% risk of female-to-male transmission). (Studies report a similar, if not more severe, risk of male-to-female transmission. See, e.g., Osmond & Padian, *AIDS Knowledge Base* 1.9-3, tbl. 1, 1.9-6 to 1.9-7.)

Second, an infected woman risks infecting her child during gestation and childbirth, i.e., perinatal transmission. Petitioner concedes that women infected with HIV face about a 25% risk of transmitting the virus to their children. 107 F.3d, at 942; 912 F. Supp., at 387, n. 6. Published reports available in 1994 confirm the accuracy of this statistic. Report of a Consensus Workshop, *Maternal Factors Involved in Mother-to-Child Transmission of HIV-1*, 5 J. *Acquired Immune Deficiency Syndromes* 1019, 1020 (1992) (collecting 13 studies placing risk between 14% and 40%, with most studies falling within the 25% to 30% range); Connor et al., "Reduction of Maternal-Infant Transmission of Human Immunodeficiency Virus Type 1 with Zidovudine Treatment," 331 *New Eng J Med* 1173, 1176 (1994) (placing risk at 25.5%); see also Strapans & Feinberg, *Medical Management of AIDS* 32 (studies report 13% to 45% risk of infection, with average of approximately 25%).

Petitioner points to evidence in the record suggesting that antiretroviral therapy can lower the risk of perinatal transmission to about 8%. App. 53; see also Connor, supra, at 1176 (8.3%); Sperling et al., "Maternal Viral Load, Zidovudine Treatment, and the Risk of Transmission of Human Immunodeficiency Virus Type 1 from Mother to Infant," 335 *New Eng J Med* 1621, 1622 (1996) (7.6%). The Solicitor General questions the relevance of the 8% figure, pointing to regulatory language requiring the substantiality of a limitation to be assessed without regard to available mitigating measures. Brief for United States as *Amicus Curiae* 18, n. 10 (citing 28 CFR pt. 36, App. B, p. 611 (1997); 29 CFR pt. 1630, App., p. 351 (1997)). We need not resolve this dispute in order to decide this case, however. It cannot be said as a matter of law that an 8% risk of transmitting a dread and fatal disease to one's child does not represent a substantial limitation on reproduction.

The Act addresses substantial limitations on major life activities, not utter inabilities. Conception and childbirth are not impossible for an HIV victim but, without doubt, are dangerous to the public health. This meets the definition of a substantial limitation. The decision to reproduce carries economic and legal consequences as well. There are added costs for antiretroviral therapy, supplemental insurance, and long-term health care for the child who must be examined and, tragic to think, treated for the infection. The laws of some States, moreover, forbid persons infected with HIV from having sex with others, regardless of consent. Iowa Code §§139.1, 139.31 (1997); Md. Health Code Ann. §18-601.1(a) (1994); Mont. Code Ann. §§50-18-101, 50-18-112 (1997); Utah Code Ann. §26-6-3.5(3) (Supp. 1997); id., §26-6-5 (1995); Wash. Rev. Code §9A.36.011(1)(b) (Supp. 1998); see also N. D. Cent. Code §12.1-20-17 (1997).

In the end, the disability definition does not turn on personal choice. When significant limitations result from the impairment, the definition is met even if the difficulties are not insurmountable. For the statistical and other reasons we have cited, of course, the limitations on reproduction may be insurmountable here. Testimony from the respondent that her HIV infection controlled her decision not to have a child is unchallenged. App. 14; 912 F. Supp., at 587; 107 F.3d, at 942. In the context of reviewing summary judgment, we must take it to be true. Fed. Rule Civ. Proc. 56(e). We agree with the District Court and the Court of Appeals that no triable issue of fact impedes a ruling on the question of statutory coverage. Respondent's HIV infection is a physical impairment which substantially limits a major life activity, as the ADA defines it. In view of our holding, we need not address the second question presented, i.e., whether HIV infection is a, *per se*, disability under the ADA.

B

Our holding is confirmed by a consistent course of agency interpretation before and after enactment of the ADA. Every agency to consider the issue under the Rehabilitation Act found statutory coverage for persons with asymptomatic HIV. . . .

[further discussion of various agencies' interpretation of disability and its application to asymptomatic HIV infection is omitted]

The regulatory authorities we cite are consistent with our holding that HIV infection, even in the so-called asymptomatic phase, is an impairment which substantially limits the major life activity of reproduction.

III

The petition for certiorari presented three other questions for review. The questions stated:

"3. When deciding under Title III of the ADA whether a private health care provider must perform invasive procedures on an infectious patient in his office, should courts defer to the health care provider's professional judgment, as long as it is reasonable in light of then-current medical knowledge?

"4. What is the proper standard of judicial review under Title III of the ADA of a private health care provider's judgment that the performance of certain invasive procedures in his office would pose a direct threat to the health or safety of others?

"5. Did petitioner, Randon Bragdon, D.M.D., raise a genuine issue of fact for trial as to whether he was warranted in his judgment that the performance of certain invasive procedures on a patient in his office would have posed a direct threat to the health or safety of others?" Pet. for Cert. i.

Of these, we granted certiorari only on question three. The question is phrased in an awkward way, for it conflates two separate inquiries. In asking whether it is appropriate to defer to petitioner's judgment, it assumes that petitioner's assessment of the objective facts was reasonable. The central premise of the question and the assumption on which it is based merit separate consideration. . . .

[material discussing whether Bragdon's health would have been endangered by treating Abbott is omitted]

We conclude the proper course is to give the Court of Appeals the opportunity to determine whether our analysis of some of the studies cited by the parties would change its conclusion that petitioner presented neither objective evidence nor a triable issue of fact on the question of risk. In remanding the case, we do not foreclose the possibility that the Court of Appeals may reach the same conclusion it did earlier. A remand will permit a full exploration of the issue through the adversary process.

The determination of the Court of Appeals that respondent's HIV infection was a disability under the ADA is affirmed. The judgment is vacated, and the case is remanded for further proceedings consistent with this opinion.

APPENDIX E

NATIONAL BIOETHICS ADVISORY COMMISSION
CLONING HUMAN BEINGS

EXECUTIVE SUMMARY (JUNE 1997)

The idea that humans might someday be cloned—created from a single somatic cell without sexual reproduction—moved further away from science fiction and closer to a genuine scientific possibility on February 23, 1997. On that date, *The Observer* broke the news that Ian Wilmut, a Scottish scientist, and his colleagues at the Roslin Institute were about to announce the successful cloning of a sheep by a new technique which had never before been fully successful in mammals. The technique involved transplanting the genetic material of an adult sheep, apparently obtained from a differentiated somatic cell, into an egg from which the nucleus had been removed. The resulting birth of the sheep, named Dolly, on July 5, 1996, was different from prior attempts to create identical offspring since Dolly contained the genetic material of only one parent, and was, therefore, a "delayed" genetic twin of a single adult sheep.

This cloning technique is an extension of research that had been ongoing for over 40 years using nuclei derived from non-human embryonic and fetal cells. The demonstration that nuclei from cells derived from an adult animal could be "reprogrammed," or that the full genetic complement of such a cell could be reactivated well into the chronological life of the cell, is what sets the results of this experiment apart from prior work. In this report the technique, first described by Wilmut, of nuclear transplantation using nuclei derived from somatic cells other than those of an embryo or fetus is referred to as "somatic cell nuclear transfer."

Within days of the published report of Dolly, President Clinton instituted a ban on federal funding related to attempts to clone human beings in this manner. In addition, the President asked the recently appointed National Bioethics Advisory Commission (NBAC) to address within ninety days the ethical and legal issues that surround the subject of cloning human beings. This provided a welcome opportunity for initiating a thoughtful analysis of the many dimensions of the issue, including a careful consideration of the potential risks and benefits. It also presented an occasion to review the current legal status of cloning and the potential constitutional challenges that might be raised if new legislation were enacted to restrict the creation of a child through somatic cell nuclear transfer cloning.

The Commission began its discussions fully recognizing that any effort in humans to transfer a somatic cell nucleus into an enucleated egg involves the creation of an embryo, with the apparent potential to be implanted in utero and developed to term. Ethical concerns surrounding issues of embryo research have recently received extensive analysis and deliberation in the United States. Indeed, federal funding for human embryo research is severely restricted, although there are few restrictions on human embryo research carried out in the private sector. Thus, under current law, the use of somatic cell nuclear transfer to create an embryo solely for research purposes is already restricted in cases involving federal funds. There are, however, no current federal regulations on the use of private funds for this purpose.

The unique prospect, vividly raised by Dolly, is the creation of a new individual genetically identical to an existing (or previously existing) person—a "delayed" genetic twin. This prospect has been the source of the overwhelming public concern about such cloning. While the creation of embryos for research purposes alone always raises serious ethical questions, the use of somatic cell nuclear transfer to create embryos raises no new issues in this respect. The unique and distinctive ethical issues raised by the use of somatic cell nuclear transfer to create children relate to, for example, serious safety concerns, individuality, family integrity, and treating children as objects. Consequently, the Commission focused its attention on the use of such techniques for the purpose of creating an embryo which would then be implanted in a woman's uterus and brought to term. It also expanded its analysis of this particular issue to encompass activities in both the public and private sector.

In its deliberations, NBAC reviewed the scientific developments which preceded the Roslin announcement, as well as those likely to follow in its path. It also considered the many moral concerns raised by the possibility that this technique could be used to clone human beings. Much of the initial reaction to this possibility was negative. Careful assessment of that response revealed fears about harms to the children who may be created in this manner, particularly psychological harms associated with a possibly diminished sense of individuality and personal autonomy. Others expressed concern about a degradation in the quality of parenting and family life.

In addition to concerns about specific harms to children, people have frequently expressed fears that the widespread practice of somatic cell nuclear transfer cloning would undermine important social values by opening the door to a form of eugenics or by tempting some to manipulate others as if they were objects instead of persons. Arrayed against these concerns are other important social values, such as protecting the widest possible sphere of personal choice, particularly in matters pertaining to procreation and child rearing, maintaining privacy and the freedom of scientific inquiry, and encouraging the possible development of new biomedical breakthroughs.

To arrive at its recommendations concerning the use of somatic cell nuclear transfer techniques to create children, NBAC also examined long-standing religious traditions that guide many citizens' responses to new technologies and found that religious positions on human cloning are pluralistic in their premises, modes of argument, and conclusions. Some religious thinkers argue that the use of somatic cell nuclear transfer cloning to create a child would be intrinsically immoral and thus could never be morally justified. Other religious thinkers contend that human cloning to create a child could be morally justified under some circumstances, but hold that it should be strictly regulated in order to prevent abuses.

The public policies recommended with respect to the creation of a child using somatic cell nuclear transfer reflect the Commission's best judgments about both the ethics of attempting such an experiment and its view of traditions regarding limitations on individual actions in the name of the common good. At present, the use of this technique to create a child would be a premature experiment that would expose the fetus and the developing child to unacceptable risks. This in itself might be sufficient to justify a prohibition on cloning human beings at this time, even if such efforts were to be characterized as the exercise of a fundamental right to attempt to procreate.

Beyond the issue of the safety of the procedure, however, NBAC found that concerns relating to the potential psychological harms to children and effects on the moral, religious, and cultural values of society merited further reflection and deliberation. Whether upon such further deliberation our nation will conclude that the use of cloning techniques to create children should be allowed or permanently banned is, for the moment, an open question. Time is an ally in this regard, allowing for the accrual of further data from animal experimentation, enabling an assessment of the prospective safety and efficacy of the procedure in humans, as well as granting a period of fuller national debate on ethical and social concerns. The Commission therefore concluded that there should be imposed a period of time in which no attempt is made to create a child using somatic cell nuclear transfer.[1]

Within this overall framework the Commission came to the following conclusions and recommendations:

I. The Commission concludes that at this time it is morally unacceptable for anyone in the public or private sector, whether in a research or clinical setting, to attempt to create a child using somatic cell nuclear transfer cloning. The Commission reached a consensus on this point because current scientific information indicates that this technique is not safe to use in humans at this point. Indeed, the Commission believes it would violate important ethical obligations were clinicians or researchers to attempt to create a child using these particular technologies, which are likely to involve unacceptable risks to the fetus and/or potential child. Moreover, in addition to safety concerns, many other serious ethical concerns have been identified, which require much more widespread and careful public deliberation before this technology may be used.

The Commission, therefore, recommends the following for immediate action:

- A continuation of the current moratorium on the use of federal funding in support of any attempt to create a child by somatic cell nuclear transfer.
- An immediate request to all firms, clinicians, investigators, and professional societies in the private and non-federally funded sectors to comply voluntarily with the intent of the federal moratorium. Professional and scientific societies should make clear that any attempt to create a child by somatic cell nuclear transfer and implantation into a woman's body would at this time be an irresponsible, unethical, and unprofessional act.

II. The Commission further recommends that:
- Federal legislation should be enacted to prohibit anyone from attempting, whether in a research or clinical setting, to create a child through somatic cell nuclear transfer cloning. It is critical, however, that such legislation include a sunset clause to ensure that Congress will review the issue after a specified time period (three to five years) in order to decide whether the prohibition continues to be needed. If state legislation is enacted, it should also contain such a sunset provision. Any such legislation or associated regulation also ought to require that at some point prior to the expiration

---

[1] The Commission also observes that the use of any other technique to create a child genetically identical to an existing (or previously existing) individual would raise many, if not all, of the same non-safety-related ethical concerns raised by the creation of a child by somatic cell nuclear transfer.

of the sunset period, an appropriate oversight body will evaluate and report on the current status of somatic cell nuclear transfer technology and on the ethical and social issues that its potential use to create human beings would raise in light of public understandings at that time.

III. The Commission also concludes that:

- Any regulatory or legislative actions undertaken to effect the foregoing prohibition on creating a child by somatic cell nuclear transfer should be carefully written so as not to interfere with other important areas of scientific research. In particular, no new regulations are required regarding the cloning of human DNA sequences and cell lines, since neither activity raises the scientific and ethical issues that arise from the attempt to create children through somatic cell nuclear transfer, and these fields of research have already provided important scientific and biomedical advances. Likewise, research on cloning animals by somatic cell nuclear transfer does not raise the issues implicated in attempting to use this technique for human cloning, and its continuation should only be subject to existing regulations regarding the humane use of animals and review by institution-based animal protection committees.
- If a legislative ban is not enacted, or if a legislative ban is ever lifted, clinical use of somatic cell nuclear transfer techniques to create a child should be preceded by research trials that are governed by the twin protections of independent review and informed consent, consistent with existing norms of human subjects protection.
- The United States Government should cooperate with other nations and international organizations to enforce any common aspects of their respective policies on the cloning of human beings.

IV. The Commission also concludes that different ethical and religious perspectives and traditions are divided on many of the important moral issues that surround any attempt to create a child using somatic cell nuclear transfer techniques. Therefore, the Commission recommends that:

- The federal government, and all interested and concerned parties, encourage widespread and continuing deliberation on these issues in order to further our understanding of the ethical and social implications of this technology and to enable society to produce appropriate long-term policies regarding this technology should the time come when present concerns about safety have been addressed.

V. Finally, because scientific knowledge is essential for all citizens to participate in a full and informed fashion in the governance of our complex society, the Commission recommends that:

Appendix E

■ Federal departments and agencies concerned with science should cooperate in seeking out and supporting opportunities to provide information and education to the public in the area of genetics, and on other developments in the biomedical sciences, especially where these affect important cultural practices, values, and beliefs.

INDEX

Index

Index